Ron S. Blicq has more than 25 years of experience in technical and business writing, and he currently conducts workshops and seminars in the United States and Canada. He is the owner of The Roning Group (communications consultants) and chairman of IEEE Professional Communication Society's Education Committee. He is the author of several articles and books, including *On the Move* and *Technically Write!*, published by Prentice-Hall.

GUIDELINES FOR REPORT WRITERS

A Complete Manual for On-the-Job Report Writing

RON S. BLICQ

A SPECTRUM BOOK

Prentice-Hall, Inc., Englewood Cliffs, New Jersey 07632

2
The Report Writer's Pyramid

If I asked you to tell me what you find most difficult about report writing, would one of these answers be yours?

> Getting started.
> Organizing the information: arranging it in the proper order.
> The writing: getting the right words down onto paper the first time.

You are in good company if your answer is similar to one of these. I can ask the same question of any group of business and technical report writers and always hear the same answers. And often those who say "Getting started" also mention one of the other answers.

The ideas presented in this book will help remove some of the drudgery from report writing. They will show you how to get started, organize your thoughts, and write simply and easily. This chapter provides you with basic guidelines. Subsequent chapters demonstrate how you can apply the guidelines to various report-writing situations.

GETTING STARTED

Frank Carter has spent two months examining his company's methods for ordering, receiving, storing, and issuing stock. He has found them to be inefficient, has investigated alternative methods, and has worked out a plan for a better system. Now he is ready to write a report describing his findings and suggestions.

But Frank is having trouble getting started. When he sits down to write, he just can't seem to find the right words. He writes a few sentences, and sometimes several paragraphs, yet each time sets them aside. He is frustrated because he feels unable to bring his message into focus.

Frank's problem is not unusual and can easily be resolved. It stems from a simple omission: he has neglected to give sufficient thought either to his reader or to the message he has to convey. What he needs to do is make two critical decisions *before* he picks up his pen. He should ask himself:

1. Who is my reader?
2. What do I *most* want to tell that reader?

Identifying the Reader

If you are writing a memo report to your manager, you will know immediately who you are writing to (although you may have to give some thought to other possible readers, if your manager is likely to circulate your memo). But if your report will have a wide readership—as Frank's may well have—then you must decide who is to be your primary reader, and write for that particular person. Trying to write for a broad range of readers can be as difficult as trying to write with no particular reader in mind. In both cases you will have no focal point for your message. And without a properly defined focal point your message may be fuzzy.

How can you identify the primary reader? It is the person (or persons) who probably will use or act upon the information you provide. You need not know the person by name, although it is useful if you do because then you will have a precise focal point. But you should at least know the type of person who will use your information and be able to identify the position he or she holds.

Yet simply knowing your reader is not enough in itself. You need to carry the identification process one step further by answering four more questions:

Question 1: What does the reader want, expect, or need to hear from me? You have to decide whether your reader will want a simple statement of facts or will expect a detailed explanation of circumstances and events. You also have to consider whether the reader needs to know how certain information was derived.

Question 2. How much does the reader know already? The answer to this question will provide you with a starting point for your report, since there is no need to repeat information the reader already knows. (But note that your answer may be influenced by the answer to question 4.)

Question 3. What effect do I want my report to have on the reader? You have to decide whether the purpose of your report is to inform or to persuade. In an informative report you simply relate the necessary facts and then stop. In a persuasive report you have to convince the reader to take some action, which can range from simply agreeing with a plan you propose, through ordering materials or equipment on your behalf, to authorizing a change in policies and procedures.

Question 4. Are other persons likely to read my report? You have to consider the route your report takes before it reaches your reader, and to whom you may send copies. If the report will pass through other persons' hands or will be carbon-copied to other persons, then you must consider how much additional information will have to be inserted to satisfy their curiosity. (At the same time you must not let your desire to satisfy these additional readers deflect you from focusing on the primary reader's needs and expectations.)

In the situation described earlier, Frank Carter decides his primary reader is John Simmonds, who is manager of Purchasing and Supply. He also recognizes that John may circulate his report to other managers, and particularly to the vice-president of the division where they all work.

Identifying the Message

Now, with his reader firmly in mind, Frank has to make his second decision. It simply means answering one question:

What do I *most* want to tell my primary reader?

Frank must examine the results of his investigation and decide which results will be most useful to John Simmonds. His aim should be to find key information which so sparks John's interest that it makes him keenly interested in learning more. For example, would John *most* want to know that

1. The company's supply system is out of date and inefficient?
2. Other businesses Frank has investigated have better supply systems?
3. There are several ways the company's supply system can be improved?
4. Improvements in the company's supply system will increase efficiency?
5. Changes in the supply system will save time and money?

Although all these points are valid, Frank reasons that John will be most interested in knowing how to save the department's time and money. But as increased efficiency is the key to these savings, Frank decides to combine points 4 and 5 into a single message. So he writes it down:

Improvements in our supply system will increase efficiency and save time and money.

This becomes Frank's *Main Message*: the information he most wants to convey to his reader, John Simmonds.

When you have identified both your primary reader and your main message, always write them in bold letters on a sheet of paper. Keep the sheet in front of you as you write, as a constant reminder that you are writing for a particular person and have a specific purpose in mind.

USING THE PYRAMID METHOD

If you were to ask any group of managers what single piece of advice they would give new report writers, the two replies you would hear more often than any other are

Tell me right away what I most need to know.

and

Draw my attention to the results. Don't bury them so I have to hunt for them.

Writers who use the "pyramid" method to organize their reports will be doing exactly what their managers want them to do. They will be emphasizing the most important information by bringing it right up front, *where it will be seen*.

As its name implies, the pyramid method suggests that you organize your reports in pyramid form, as shown in Figure 2-1. The essential information (what the reader most needs to know) sits at the top of the pyramid, and is supported by a strong base of facts and details.

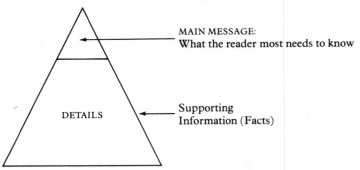

Fig. 2-1. The basic pyramid.

The concept is not new. It has been used for decades by journalists, whom we often call "reporters" because, like us, *they write reports*. And in recent years experienced business and technical report writers have adopted it, because it offers them the most efficient way to communicate information. For writers in government, business and industry, however, the two compartments which form the pyramid are relabeled slightly, as shown in Figure 2-2.

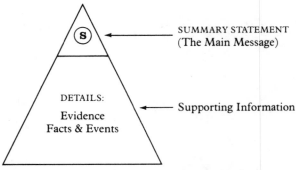

Fig. 2-2. The report writer's pyramid.

This pyramid is used as the basic framework for organizing every type of report, even though the compartments may be relabeled and expanded to suit varying situations. I want you to become so accustomed to using it that you automatically think "pyramid style" whenever you have to write a report.

Focusing the Message

Many report writers find the pyramid method difficult to accept at first, because it seems to contradict what they have previously been taught. Throughout high school, college, or university they have probably been told to write using the "climactic" method. Writing climactically means developing a topic carefully, logically, and sequentially, so that the narrative leads systematically up to the main point. It is the ideal way to write an essay, short story, or mystery novel, in which the main point needs to be at the end of the piece of writing. But it does not meet the needs of business and industry, where readers want to find the Main Message at the beginning.

This does not mean that you have to completely discard the climacttic method of writing: it can still be an effective way to write the Details section of a report. The pyramid method simply suggests that you identify the most important information in the Details section, and then condense it into a short Summary Statement which you place at the front of your report. In this way you focus your readers' attention on your Main Message.

If you have been accustomed to writing climactically, you may feel uncomfortable jumping straight into the Main Message without first leading gently up to it. To help you get started I recommend borrowing a technique used by newspaper reporters.

Turn to the front page of your daily newspaper and read the first few paragraphs of each article. You will find that every article is structured the same way:

1. It has a headline, which really is not part of the article (headlines are normally written by editors, not the newspaper reporters who write the articles.)
2. Its opening paragraph very briefly gives you the main information—usually what has happened and, sometimes, the outcome. For example:

    Maps distributed in Washington yesterday show that the 77.5-ton Skylab space station will pass directly over Denver during its final days in orbit. It is expected to enter the earth's atmosphere on Wednesday, July 11, plus or minus one day.

    This opening paragraph is the article's "Main Message," and is equivalent to the Summary Statement at the front of a short report.
3. Its remaining paragraphs expand the Main Message by providing details, such as facts, events, names of places and people, dates and times, and statements by persons the reporter has interviewed. It is equivalent to the Details section of a report.

What you cannot see are six "hidden words" at the front of each article, which newspaper reporters use every time they start writing. First they write

I want to tell you that...

And then they finish the sentence with their Main Message (what they *most* want to tell their readers). For example:

> I want to tell you that . . . there are larvae in the city's water supply, but local authorities say they don't pose a threat to public health.

Finally, reporters *remove* the six words, "*I want to tell you that . . .*" (which is why they are known as "hidden words"), so that the remaining words become the article's opening sentence.

You can see how this is done if I restore the six hidden words to the front of the Skylab opening paragraph quoted earlier:

> *I want to tell you that* . . . maps distributed in Washington yesterday show that the 77.5-ton Skylab space station will pass directly over Denver during its final days in orbit.

Similarly, if you return to the front page of your daily newspaper you should find that you can also insert the six hidden words before the opening sentence of every article you read.

You can use the hidden words technique to help you start every report you write. The steps are simple:

1. Identify your reader.
2. Decide what you *most* want to tell your reader.
3. Write down the six words "I want to tell you that. . . ."
4. Complete the sentence by writing what you have decided to tell your reader (from step 2). This is your Main Message.
5. Delete the six "hidden words" of step 3.

Here is how Frank Carter used these five steps to start his report on the company's supply system:

1. He identified his reader as John Simmonds, manager of Purchasing and Supply.
2. He decided he wanted to tell John that the department needs to improve its supply system.
3. He wrote: "I want to tell you that. . . ."
4. He finished the sentence: ". . . improvements to our supply system will increase efficiency and save time and money."
5. He deleted the six words he had written in step 3.

Step 4 became the opening sentence of Frank's report, that is, his Summary Statement (or Main Message). But when Frank examined the words more closely he realized that although what he had written was accurate, as an opening statement it was too abrupt. He remembered that a Summary Statement must not only inform but also create interest, and thus encourage the reader to continue reading. So he rearranged his information and inserted additional words to soften the abruptness. At the same time he took great care not to lose sight of his originally defined Main Message. After several attempts, this is what he wrote:

My examination of our supply system shows we can increase departmental efficiency, save time, and reduce costs by improving our methods for ordering, storing, and issuing stock.

I suggest that you, like Frank Carter, use the "hidden words" method every time you have to write a report. It will help you start more easily and ensure that you focus your readers' attention immediately onto the most important information.

Developing the Details

Because the Summary Statement of a report brings readers face to face with important, sometimes critical, and occasionally controversial information, it immediately triggers questions in their minds. It is your responsibility to anticipate these questions and answer them as quickly and efficiently as you can. This is done in the Details section, which amplifies and provides supportive evidence for the Main Message in your Summary Statement.

There are six questions a reader may ask: *Who?*, *Why?*, *Where?*, *When?*, *What?*, and *How?* (see Figure 2-3). But first you have to identify which of these questions your reader would be likely to ask. You say to yourself: "If I were the intended reader, which questions would I ask after I had read only the Summary Statement?" (You also have to remember that your readers won't know the subject nearly as well as you do.) And then you select the appropriate questions from Figure 2-3.

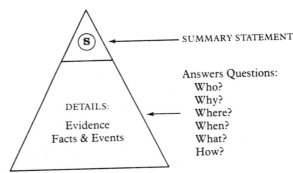

Fig. 2-3. The base of the report writer's pyramid answers readers' questions.

Frank Carter, for example, would say to himself: "What questions will John Simmonds be likely to ask immediately after he has read my Summary Statement?" (It's in bold type, above.) And Frank would probably answer his own question with

- **Why** (is it necessary to increase efficiency)?
- **How** (can we improve efficiency)?
- **What** (will be the effect or result of improved efficiency)?

Frank might also answer *when?* and *who?*, depending on whether he thinks John Simmonds wants a very detailed report. The questions would then be:

- **When** (should the improvements be implemented)?
- **Who** (will be affected by them)?

(He would omit the question *where?*, because for this report it does not need to be answered.)

Now let's examine how Bev Hunter used these six questions to develop the Details section of a short inspection report. Recently Bev drove to a warehouse to determine the condition of some new equipment damaged in a traffic accident, and found that most of it was beyond repair. In the Summary Statement of the report Bev told readers what they most needed to know (the *Main Message*):

> (I want to tell you that . . .) Our inspection shows that only three of the 16 electronic typewriters in Calvin Computer Systems shipment No. 367 can be repaired. The remainder will have to be scrapped.

To assemble facts for the Details section, Bev jotted down notes in answer to the questions readers might ask after they had read the Summary Statement:

> ***Who*** *(was involved)?*
>   Fran Derwood and Bev Hunter.
> ***Why*** *(were you involved)?*
>   We had to inspect damaged typewriters.
>   (Authority: Arlington Insurance Corporation.)
> ***Where*** *(did you go)?*
>   To Hillsboro Storage warehouse.
> ***When*** *(did this happen)?*
>   On June 13.
> ***What*** *(did you find out)?*
>   3 repairable typewriters; 13 damaged beyond repair.
> ***How*** *(were they damaged)?*
>   In a semitrailer involved in a highway accident.

Finally, Bev took these bare facts and shaped and expanded them into two Details paragraphs:

Details:

Why?

How?

Who?
Where?
When?

What?

> We were requested by Arlington Insurance Corporation to examine the condition of 16 CANFRED electronic typewriters manufactured by Calvin Computer Company, Montrose, Ohio. They were damaged when the semitrailer in which they were shipped overturned and burned on a curve near Hillsboro, Maryland, on June 11. Fran Derwood and I drove to Hillsboro on June 13, where we were met by Arlington Insurance Corporation representative Kevin Cairns, who escorted us to the Hillsboro Storage warehouse.
>
> We found that the fire which resulted from the accident has irreparably damaged 13 typewriters. Three other typewriters suffered smoke and heat damage, but seem to be mechanically and electrically sound; they carry serial numbers 106287, 106291, and 106294. We estimate that these typewriters will cost an average $350 each to repair, for a total repair cost of $1050.

The pyramid method can help you organize random bits of information, just as it has helped Bev Hunter. And because it helps you eliminate nonessential information, it will also shorten the length of reports you write. But it is not meant to be a rigid method for organizing details. The six basic questions are intended solely as a guide, and should be used flexibly. For example:

- The questions do not have to be answered in any particular order. You can arrange the answers in any sequence you like, balancing your personal preference against the reader's needs and the most suitable way to present your information.
- Only the appropriate questions need be answered (i.e. the questions that are pertinent to each particular reporting situation).
- The first four questions in the list (*who?*, *why?*, *where?*, and *when?*) require fairly straightforward answers. The last two questions (*what?* and *how?*) can have widely varying answers, depending on the event or situation you are reporting. This is where you explain what has happened, what needs to be done, and possibly how best to go about it. Consequently there is ample scope for originality and ingenuity on your part.

Expanding the Details Section

The pyramid method provides the basic structure for all reports, regardless of their length. It can be used for one-paragraph reports, one-page reports, and 100-page reports. In every case the reports open with a **Summary** (or **Summary Statement**), which presents a synopsis of the main information to be conveyed (this is the "Main Message"). It is followed by a longer section containing factual Details, which support and amplify the initial statement.

Bev Hunter's report describing the inspection of damaged electronic typewriters is a typical short report structured according to the simple two-part pyramid. But for reports of greater length or complexity, the Details section at the base of the pyramid needs to be developed further.

This development is obtained by expanding the Details section into three basic compartments of information:

- A **Background** compartment, which describes the circumstances leading up to the situation or event. (It answers the first four questions, *who?*, *why?*, *where?*, and *when?*)
- A **Facts and Events** compartment, which describes in detail what happened, or what you found out during your project. (It answers the last two questions, *what?* and *how?*)
- An **Outcome** compartment, which describes the results of the event or project, and sometimes suggests what action needs to be taken. (It also can answer the questions *what?* and *how?*)

The compartments are shown within the pyramid in Figure 2-4, and are keyed to the appropriate parts of the short report in Figure 2-5.

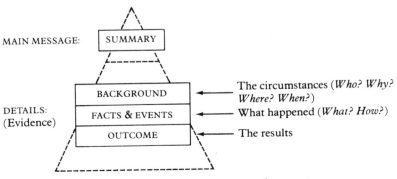

Fig. 2-4. The main compartments of a report.

SUMMARY

BACKGROUND
Answers
questions:
    Who?
    When?
    Why?

FACTS & EVENTS
Answers
questions:
    How?
    What?

OUTCOME
Answers
questions:
    How?
    What?

To: D. Shwenck, Marketing Manager, Head Office
From: Carl Peebles, Manager, Store No. 6

(I want to tell you that . . .) The installation instructions included with Vancourt BK-7 door locksets contain errors which should be corrected before any more locksets are sold.

The Vancourt BK-7 is a newly designed lockset which replaces the popular BK-6. We received our first shipment on October 12 and had sold nine of them by October 19 before we realized they were apparently faulty. By then, six customers had returned their new locksets with the complaint that they could not be installed.

I tried installing one of the returned locksets and discovered that the drilling template provided with the kit is inaccurate. There are two errors:

1. The hole for inserting the lockset through the door should be drilled with a 2⅝ in. diameter, not the 1⅝ in. diameter scribed on the template.
2. The hole should be centered 2⅞ in. from the door edge, not the specified 2⅛ in.

We have drawn a new template (see attachment) and have inserted a copy into every BK-7 lockset kit in our stock. We have also telephoned Vancourt Manufacturing Company to inform them of the error. I suggest that our other stores should be warned of this error and told how to correct the instructions supplied with any BK-7 locksets they have in stock.

Fig. 2-5. A short report structured in the pyramid style. The main report writing compartments are identified on the left side of the report.

These four compartments—

SUMMARY
BACKGROUND
FACTS AND EVENTS
OUTCOME

—provide the basic framework for every report you are likely to write. You will be able to identify them in every report described in Chapters 3 through 8, although often you will find the compartments relabeled to suit particular situations. In longer reports the compartments are also subdivided to accommodate a greater bulk of information and to improve internal organization. These subdivisions occur mostly in the Facts and Events compartment.

II
INFORMAL REPORTS

3
Occurrence, Field Trip, and Inspection Reports

The reports in this chapter are normally short, each containing one to three pages of narrative and occasionally attachments such as drawings, photographs, and calculations. They are all structured using the report writer's pyramid described in Chapter 2, with the pyramid modified slightly and relabeled to suit various situations.

For each type of report described here and in subsequent chapters, the guidelines comprise

1. A description of the report, an illustration of its pyramid structure, and definitions of each writing compartment.
2. One or two examples of a typical report.
3. Comments on how the particular type of report should be or has been written, with cross-references to the example.

Short reports such as these are most often written in memorandum form, sometimes as letters, and occasionally as semiformal reports with a title at the top of the page. For examples of typical formats, see Chapter 9.

OCCURRENCE REPORTS

An Occurrence Report (sometimes referred to as an Incident Report) tells about an event that has happened, explains how and why it occurred, and describes what effect the event had and what has been done about it. Sometimes it also suggests that corrective action be taken, or what should be done to prevent the event from recurring.

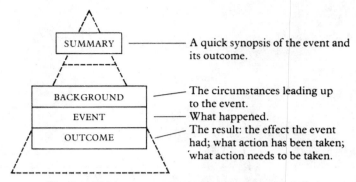

Fig. 3-1. Writing compartments for an occurence report.

The writing compartments are similar to those of the basic report writer's pyramid, and are shown in Figure 3-1. In these compartments,

- The **Background** answers the questions Who?, Why?, Where?, and When?
- The **Event** and **Outcome** answer the questions How? and What?

The depth of detail provided in each compartment depends on the importance of the event and how much the reader wants or needs to know about it. For example, if you are informing your comptroller of an accounting error that caused a supplier to be overpaid $100, you would write just a brief report. But if you are describing an accident which hospitalized two employees and cost $30,000 in repair work, you would be expected to write a detailed report that fully describes the circumstances and the corrective action that has been taken.

The comments that follow identify the four writing compartments in a Short Occurrence Report.

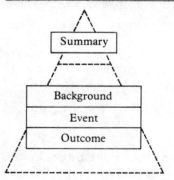

OCCURRENCE REPORT:
Reporting a Pricing Error

In his occurrence report, Paul Willis, an assistant manager at the Provo Department Store, is describing a pricing error to Jim Knox, his manager.

1. In his **Summary Statement** Paul explains very briefly what happened and the effect it had. (Note that the hidden words "I want to tell you that . . ." can be placed not only in front of this paragraph, but in front of *all* paragraphs. This is not unusual for a short report.)

2. In his **Background** compartment Paul answers the four questions his reader (Jim Knox) is likely to ask:

   What barbecues were they? (Cranston CBQ)
   When did this happen? (Today)
   How did it happen? (Details of error)
   Who is responsible? ("I": Paul)

Note how the words "They were . . ." build a successful transition between Paul's Summary Statement and the beginning of the Background compartment.

3. The **Event** compartment describes exactly what happened and explains why the event could not have been averted (it answers the question *Why?*).

4. In his **Outcome** compartment Paul answers the question *What* (did you do about it?).

Paul has a choice regarding the position of the last sentence of paragraph 2, in which he explains how the printing error occurred. Alternatively, he could include it as part of paragraph 4 (Outcome), on the assumption that it is information he obtained after the event.

As a report writer you will sometimes encounter situations like this, in which you have to decide where to place a particular piece of information. You should base your choice on which feels most comfortable from both your and the report reader's viewpoint.

MEMORANDUM

From: Paul Willis   Date: March 31, 19xx

To: J Knox, Manager   Subject: Pricing Error
    Electrical Appliances

(1) An advertising error has forced us to sell 30 electric barbecues at $27.00 below the planned sale price, for a total loss of $810.00.

(2) They were Cranston model CBQ barbecues which are regularly sold for $129.95. They were to be reduced to $96.95 for today's personal shopping Yellow Tag sale. However, last night's newspaper advertisement displayed a sale price of only $69.95. The error occurred during typesetting, and I did not notice it when I OK'd the proof yesterday afternoon.

(3) All 30 barbecues on the floor had to be sold at the advertised price because they were claimed by customers within two minutes of door opening, before we were aware of the error.

(4) I immediately posted a "sold out" sign, and have inserted a price correction in this afternoon's newspaper.

Paul

TRIP REPORTS

Trip reports are written whenever people leave their normal place of work to do something elsewhere. Their reports can cover many kinds of events, such as

- Installation or modification of equipment
- Assistance on a field project
- Attendance at a conference, seminar, or workshop
- Repairs to a client's equipment or field instruments
- Evaluation of another firm's buildings, facilities, or methods

Whatever the circumstances, the writing compartments for a field trip report are essentially the same as those for the basic report writing pyramid described in Chapter 2. The "Event" compartment, however, is relabeled THE JOB, as shown in Figure 3-2.

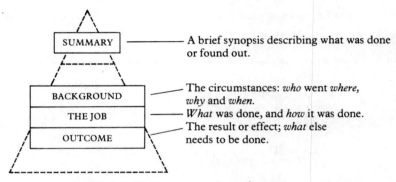

Fig. 3-2. Writing compartments for a trip report.

These compartments generally contain the following information:

- The **Background** compartment describes the purpose of the trip, mentions on whose authority it was taken, and lists circumstantial details such as the names of persons involved, and dates and locations.
- **The Job** compartment describes what was done. Often, it can be broken down into four subcompartments:
  1. What the report writer set out to do
  2. What was actually done
  3. What could not be done, and why
  4. What else was done

  The fourth subcompartment is necessary because people on field trips often find themselves doing things beyond the purpose of their assignment. For example, a technician sent in to repair a defective diesel power unit at a remote radar site may notice that a second unit is "running rough," and spend an additional six hours adjusting its timing cycle. The time spent on this additional work must be accounted for, and the work must be described in his or her trip report.
- The **Outcome** compartment sums up the results of the trip and, if further work still needs to be done or follow-up action should be taken, suggests what is necessary and even how and by whom it should be accomplished.

Many trip reports are short and simply follow the compartment arrangement in a few paragraphs. Some longer, more detailed trip reports may need headings to break up the narrative into visible compartments. Typical headings might be

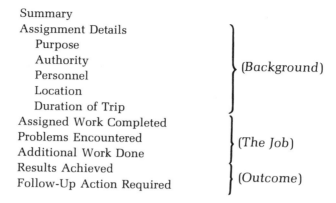

```
Summary
Assignment Details    ⎫
    Purpose           ⎪
    Authority         ⎬ (Background)
    Personnel         ⎪
    Location          ⎪
    Duration of Trip  ⎭
Assigned Work Completed ⎫
Problems Encountered    ⎬ (The Job)
Additional Work Done    ⎭
Results Achieved        ⎫
Follow-Up Action Required ⎬ (Outcome)
```

Trip reports are often submitted as memorandums, written by the person who made the trip (or who was in charge of a group of people who did the job as a team), and addressed to his or her supervisor or manager. It is natural, therefore, to use the first person for such reports ("I" or "we"). This personalization of short memorandums or letter reports is encouraged, because it helps report writers sound confident. The first person is used in both of the sample trip reports that follow, which are

A report on a field installation
A report on seminar attendance

A method for adapting trip report compartments so they can be used to report conference attendance follows the second trip report.

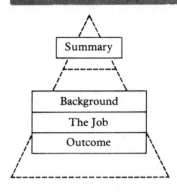

TRIP REPORT NO. 1:
Reporting an Installation

Frank Crane is a field service representative for Vancourt Business Systems Inc., and is reporting to his company's R & D manager, Dale Rogerson, on an installation he has just completed.

1. For his **Summary**, Frank has picked information primarily from the Background and The Job compartments.

2. The **Background** compartment brings together all the bits and pieces of information relating to the trip. (It answers the question *Who went where, why, and when?*)

    Note that Frank has spent some time describing the purpose of the modification kit, so that readers not familiar with it will be better able to understand his report.

3. **The Job** compartment starts here.

    For a trip report describing an installation, modification, or repair work, follow these guidelines:

    a. Describe routine work that goes according to plan as briefly as possible, particularly if there is an instruction or work specification which can be referred to and/or attached.
    b. Describe unexpected work, unusual events, or problems in some detail, and
4. particularly explain how a difficult situation has been resolved.

5. To conserve space the diagram and invoice have been omitted.

6. Sometimes it can be difficult to identify exactly where the Outcome compartment starts. Some people might say it starts at the beginning of the previous paragraph; others might say it starts at this paragraph; while still others might argue it starts at the next paragraph.

    But remember that only Frank really needs to know where the Outcome starts (because he uses it to organize his report). All that is necessary from the reader's point of view is for the report to read smoothly and progress logically from beginning to end.

7. In this final **Outcome** statement, Frank provides a memory jogger for his readers. In effect he is saying to both Dale Rogerson (to whom the report is addressed) and Jerry Morganski (who receives a copy): "Take note! Plan to send someone to Westland between January 20 and January 24."

VANCOURT BUSINESS SYSTEMS INC

From: Frank Crane, Field Service Rep. Date: October 23, 19xx

To: Dale Rogerson, Manager Subject: Installation of Prototype
Research and Development Modification Kit MCR-1

① An MCR-1 multi-account readout display and control box have been installed on a model 261 Processor, where they will be field-tested for three months.

② I was assigned by Work Order M97 to install the prototype kit on a processor owned by Arrow Industries at Westland, where arrangements had been made for it to be field-evaluated. Modification kit MCR-1 permits raw data on individual accounts stored in Vancourt 261 Processors to be made instantly available on a miniature display unit mounted beside the processor. I drove to Westland on October 19 and returned on October 22.

③ The circuit and control box were installed without difficulty. However, a locally manufactured equipment rack on which the 261 Processor has been mounted prevented installation of the miniature display unit beside the processor, as directed in step 29 of the installation specification.

④
⑤ I arranged for the mounting tray to be modified by Corwin Metals in Westland, so that it could be mounted on top of the processor as shown on the attached diagram. Corwin Metals' invoice for $116.25 is attached.

⑥ I tested the control box and miniature display unit, and detected and corrected two display faults. I then tested the installation for three more hours, but detected no further faults. During this period I trained three employees of Arrow Industries to use the equipment.

Their contact employee will be Lorne Carter, who is one of the three persons I trained, and with whom I left evaluation and serviceability status report forms. He will mail these to you weekly.

⑦ Arrangements will have to be made to disconnect and remove the kit during the week of January 20 to 24, 19xx.

Frank Crane
FC:jp

cc Jerry Morganski, Field Service Manager

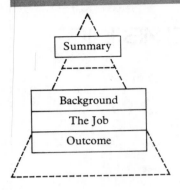

TRIP REPORT NO. 2:
Reporting Seminar Attendance

Personnel assistant Karen Young has recently attended a seminar and is evaluating it in a report to personnel manager Audrey Rivers.

1. This subject line is effective because it contains a word which describes what Karen was *doing*. (She writes: "Evaluation of . . ." [the seminar]; if she had written "Attendance at . . .," it would not have been as informative.)

2. Karen's enthusiastic **Summary** is drawn primarily from the Outcome compartment (see 5).

3. This **Background** compartment does three things:

    a. It describes the seminar.
    b. It outlines Karen's involvement in it.
    c. It answers the questions Who?, Why?, Where?, and When?

4. Karen divides **The Job** compartment into two subcompartments:

   4A   a. A brief description of how the seminar was organized and, by implication, what she did as a seminar participant
   4B.   b. An evaluation of the seminar's effectiveness and usefulness

5. For this type of report, the **Outcome** compartment is particularly important because it provides an answer to the question, Was the purpose of the trip achieved? Karen's manager wants to know how good the seminar is, and Karen has told her confidently that the company should buy the in-house version of it, and has even suggested that she could be workshop leader for it.

MEMORANDUM

To: Audrey Rivers
Personnel Manager

From: Karen Young
Personnel Assistant

Date: March 19, 19xx

Subject: Evaluation of Roning Group "Meetings" Seminar

The Roning Group "Meetings" seminar is thoroughly worth attending and should be made available to all supervisory employees.

The attached folder describes the seminar in detail. Its full title is "Increasing the Effectiveness of Business Meetings" and it is available in two formats: as a public seminar such as the one I attended, and as a kit for in-house presentation. Attendance at a public seminar costs $35.00 per person. The kit, which contains sufficient materials to train 48 employees, costs $600.00, which is equivalent to $12.50 per person.

The seminar I attended was held at the downtown Holiday Inn from 1:00 to 4:30 p.m. yesterday, March 18. I was selected to attend so that I could evaluate its suitability for in-house use.

The seminar was divided into three one-hour compartments:

1. For the first hour the seminar leader demonstrated techniques for improving meeting performance, and then we prepared to take part in two meetings. We were divided into two groups, and we were each given two roles to play.

2. During the second hour the first group held a meeting, and the second group evaluated performance on a one-to-one basis.

3. During the third hour the groups' roles were reversed. The seminar leader also summed up after each meeting and pointed out our strengths and weaknesses.

I found the seminar to be an excellent learning experience. The atmosphere was relaxed, the participants were actively involved for more than 75% of the time and had realistic, recognizable roles to play, and the topics for the two meetings were both interesting and relevant. Of particular value were the comments we could make on one another's performance as meeting participants.

I suggest that we purchase The Roning Group's "Meetings" kit and use it to help our managers and supervisors hold more efficient meetings. The experience I gained at yesterday's seminar has prepared me sufficiently to run in-house workshops.

Karen

Reporting Conference Attendance

The trip report compartments can be used to describe attendance at a conference or meeting. The most difficult compartment to write is THE JOB, and the most efficient way to organize it is to divide it into subcompartments such as these:

- What the person attending the conference expected to gain, learn, or find out.
- What the program promised would be covered.
- What sessions were attended, and why they were chosen. (Important for a conference with several simultaneous sessions.)
- What was gained or learned by attending these sessions.
- What was gained or learned from meeting and talking to other persons attending the conference.
- What other activities were attended.

INSPECTION REPORTS

An inspection report is similar to a field trip report, in that its writer has usually gone somewhere to inspect something. Bev Hunter's trip to Hillsboro, Maryland, to inspect damaged electronic typewriters is an example (see Chapter 2).

Other typical situations requiring an inspection report to be written are:

- Examination of a building to determine its suitability as a storage facility.
- Inspection of construction work, such as a bridge, building, or road.
- Checks of manufactured items, to assure they are of the required quality.
- Inspection of goods ordered for a job, to check that the necessary quantities and correct items have been received.

The writing compartments are also similar to those for a trip report, except that the compartment previously labeled simply "The Job" can be more clearly defined as "Findings." The report writer's pyramid is illustrated in Figure 3-3.

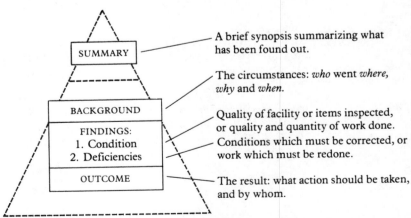

Fig. 3-3. Writing compartments for an inspection report.

27 Occurrence, Field Trip, and Inspection Reports

The following notes suggest what these compartments should contain and how the information should be arranged:

- The **Background** compartment describes the purpose of the inspection, mentions on whose authority it was performed, and lists circumstantial details such as the names of persons involved and the date and location of the inspection.
- The **Findings** compartment often can be divided into two subcompartments, one of which describes *conditions found*, while the other lists *deficiencies* (a deficiency can be either an unacceptable condition or a missing item). The length and complexity of the Findings will dictate how these compartments are organized:

Short, simple Findings can be arranged in the order suggested:

Conditions Found:
    1. -------
    2. -------
    3. ------- (etc.)
Deficiencies:
    1. -------
    2. ------- (etc.)

This arrangement is illustrated in Inspection Report 1 (items 4 and 5).

Longer, more complex Findings should be arranged so that the deficiencies for each item are listed immediately after the item's condition has been described, like this:

Inspection Findings:
Item A
    Condition
    Deficiencies
Item B
    Condition
    Deficiencies
        (etc.)

This arrangement is shown in Inspection Report No. 2 [see items 4, 5, and 6]. The intent is to keep the deficiencies reasonably close to the condition from which they evolve.

- The **Outcome** compartment suggests what should be done as a result of the inspection. If deficiencies have been listed at the end of the previous compartment, the Outcome is likely to be short:

    Provided that the deficiencies I have listed are corrected, the warehouse should make a suitable storage facility for the Passant Project.

The two inspection reports on the following pages demonstrate how these guidelines are applied. Notice particularly how the use of the first person ("I") is employed by both writers, but it occurs more often and more naturally in the report prepared as a memorandum (report No. 2).

A basic form for recording inspection information is shown in Figure 3-4.

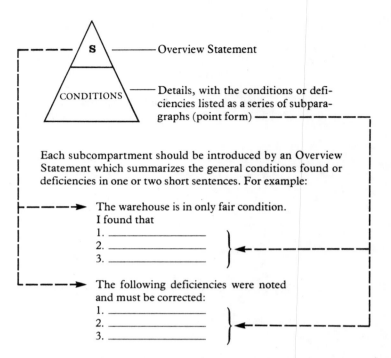

Each subcompartment should be introduced by an Overview Statement which summarizes the general conditions found or deficiencies in one or two short sentences. For example:

The warehouse is in only fair condition. I found that
1. _____
2. _____
3. _____

The following deficiencies were noted and must be corrected:
1. _____
2. _____
3. _____

It's also a good plan to use a miniature pyramid to organize the information in the *Conditions Found* and *Deficiencies* subcompartments (above).

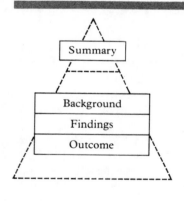

INSPECTION REPORT NO. 1: Inspecting a Contractor's Work

1. Paul Thorvaldson, the author of this report, has chosen to use a slightly more formal format than a memorandum would offer. He reasons that several people will read the report, and the contractor will likely be given a copy when asked to rectify the deficiencies.

2. The **Summary** tells readers right away what they most want to know: that the new facility is ready.

3. In the **Background** compartment, Paul describes the details leading up to his inspection visit. He answers the questions Who?, Where?, When?, and Why?

4. The **Findings** start here with an *overview statement* ("The contractor has done a good job"). They continue immediately with *Conditions Found*, which Paul limits to two short sentences mentioning the main items he noticed.

> When specifications have been met, there is no need to describe everything that has been done; it is sufficient simply to indicate that the job has been completed correctly. But when specifications have not been met, attention must be drawn to every item which has been improperly done (see 5).

5. The **Findings** continue with the **Deficiencies**. Each item needing correction is listed in a separate subparagraph (to make it easy to identify step-by-step what action has to be taken), and in clearcut terms which will not be misunderstood. If there are many deficiencies it may be more convenient to list them on a separate sheet or sheets (called an attachment), and to refer to them in the Deficiency paragraph, like this:

> The 27 deficiencies listed on the attached sheets must be corrected by the contractor before the job can be accepted.

6. The **Outcome** compartment describes the result of the inspection (in this case, whether the contractor's work can be accepted and the new accommodation occupied). The Outcome provided Paul with the primary information he needed for the Summary at the start of his report.

---

robertson
engineering
company

INSPECTION REPORT

ALTERATIONS TO AND REDECORATION OF NEW OFFICES
FOR THE ENVIRONMENT AND RESOURCES DEPARTMENT

①

② Except for some minor deficiencies, the repair and renovation work is complete. The offices can be occupied on November 1, as scheduled.

③ These offices were previously occupied by Nor-West Distributors, who vacated them on September 30. A contract for renovating and redecorating the premises was let to Craven Builders Inc on September 25; it specified the work to be done and an October 30 completion date. The contractor notified us on October 28 that the work was complete, and I inspected the offices on the morning of October 29.

④ The contractor has done a good job. There are no signs that the previous temporary walls have been removed or existed, and the new temporary walls look like permanent structures. Decorating quality is very good.

⑤ The following deficiencies were noted and must be corrected by the contractor:

1. The rubber underlay needs to be relaid under parts of the fitted carpet, particularly in the northeast corner of the main office; currently it is bunched in seven or eight places.

2. The drapery tracks over each window need to be extended to the specified width of 6 feet; currently they are only 4 feet wide.

3. The door to the manager's office needs to be rehung so that it closes fully and the lock engages.

4. Four of the lighting fixtures need to be realigned so that they are all at the same level.

5. The cove needs to be installed at the foot of all walls.

These deficiencies will not prevent immediate occupation of the new offices, although I suggest that furniture should not be placed in the northeast section of the main office until the carpet underlay has been relaid. All other deficiencies can be corrected without interfering with the department's operations.

I suggest that a second inspection be scheduled before contractor payment is approved.

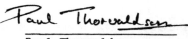

Paul Thorvaldson
October 29, 19xx

INSPECTION REPORT NO. 2:
Inspecting Printing Equipment

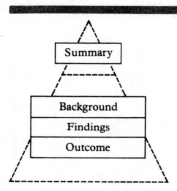

Fred Parkin has been examining some used printing equipment his company is considering buying, and reports his findings to comptroller Keith Adams.

1. This is Fred's **Summary**. He answers the comptroller's immediate question: Should they buy the printing equipment?

2. In the **Background** compartment Fred answers the questions Who?, Where?, When?, and Why? He gives more detail than the comptroller probably needs because he is aware that other management persons who may be involved in the decision may not know the circumstances.

3. The **Findings** compartment starts with a comment on the equipment's overall condition. An *overview statement* like this prepares the reader for the specific details that follow. The Findings continue through the remainder of the report, except for the final paragraph.

4. Fred has chosen to treat each major piece of equipment separately, and to comment on both its Condition and Deficiencies under the one heading.

5A. to 5D. These are the *Conditions Found* for each major piece of equipment. The comptroller would not expect Fred to give a detailed description of each equipment's condition.

6A. to 6D. These are the major *Deficiencies* for each piece of equipment. They are not listed as specific, item-by-item deficiencies because it would have taken much time and effort for Fred to prepare a detailed list. To do so would seem particularly pointless when Fred is suggesting that they not buy the equipment.

Occurrence, Field Trip, and Inspection Reports

6B. This is an unusual "deficiency." Rather than state what action would have to be taken, it comments on the suitability of the equipment compared to more modern systems.

7. This is the **Outcome** compartment, which sums up Fred's feelings about the proposed purchase. It was from this paragraph that Fred drew most of the information for his Summary (at 1).

H. L. Winman and Associates

INTER - OFFICE MEMORANDUM

From: Fred Parkin Date: May 20, 19xx

To: Keith Adams, Comptroller Subject: Inspection of Cam Industries' Printing Equipment

(1) Although Cam Industries is selling its printing equipment at a very attractive price, I would hesitate to buy any of it.

(2) As requested in your memo of May 17, I visited Cam Industries at 2820 Wampole Road this morning to inspect their printing facility. They will be closing their in-house print shop on June 30 and will be selling all the equipment for $15,000, or individual items for a total of $17,500. I was to examine the equipment and consider whether all of it, or individual items, would make a practicable purchase for the facility we plan to set up to print parts lists and brochures.

(3) I found the printing equipment to be in only fair condition. Most of it has been in use for more than 12 years, and seems to have been operated by untrained help. The condition of major items is outlined briefly below:

(4) Bellweather 316 Offset Duplicator

(5A) This unit is 15 years old but looks like 20! Service has been erratic, with service reps being called in only upon equipment failure (the present owners apparently do not believe in preventive maintenance).

(6A) I estimate that almost all moving parts will have to be replaced immediately or very shortly, for a total repair cost of $6000. This was confirmed by the local Bellweather service rep, who knows the equipment.

2/ ...

Flexite 60 Camera and Mounting Frame

(5B) (6B) The unit is 12 years old and is in generally good condition. It has no moving parts, so no noticeable wear has taken place. The camera bellows seem to be satisfactory but probably should be inspected professionally. At $1600 the unit is an excellent buy. But it remains an antiquated, slow system to use when compared to newer systems which are more efficient and have much greater flexibility.

Selenium Plates (12) and Baking Oven

(5C) Both are in poor condition. The plates are badly scratched and the oven has numerous heat cracks and shows signs of leaks.

(6C) Both need to be replaced at a cost of $3800.

Collator, Binders, Staplers, Padders, etc

(5D) (6D) There is a varied assortment of equipment in varying condition, with a total sale value of $900. About half could be retained, but the other half would have to be repaired or replaced at a cost of $1600.

(7) If we purchase Cam Industries' printing equipment for $15,000, we will almost immediately have to invest a further $11,400 for repairs and replacement items, and even then we will own outdated equipment which is at least 12 years old. I believe this would be an uneconomical purchase.

F. P.

INSPECTION REPORT

| Location: _____ Date: _____ |
| Item(s) being inspected: |
| Inspector: _____ Contractor: _____ |
| CONDITIONS FOUND: |
| DEFICIENCIES: |
| RECOMMENDATION(S): |

Fig. 3-4. A form for an inspection report.

4

Progress Reports and Short Investigation Reports

Like the reports in Chapter 3, these informal reports are short, seldom exceeding three pages plus attachments. Their writing compartments, however, are often expanded to include more subdivisions, particularly in the FACTS AND EVENTS compartment of the basic pyramid (Figure 2-4).

When a report contains detailed information, such as lists of materials, cost analyses, schedules, or drawings, they are normally removed from the body of the report (from the FACTS AND EVENTS compartment) and placed at the back, where they are referred to as "Attachments" or "Appendices." (This is done to avoid cluttering the report narrative with tabular data and thus interrupting reading continuity.) As they provide supportive evidence, or "backup," for statements made in the report, a separate compartment is created for them at the foot of the report writer's pyramid. This compartment is labeled BACKUP and is shown with a dotted line in Figure 4-1 to indicate that it is optional.

PROGRESS REPORTS Progress reports keep management informed of work progress on projects that span a lengthy period, which can vary from a few weeks for a small manufacturing contract to several years for construction of a hydroelectric power station and transmission system.

There are two types of progress reports:

Occasional Progress Reports are written at random intervals and usually concern shorter projects.

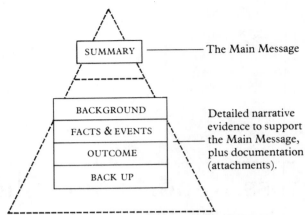

Fig. 4-1. The report writer's pyramid with a "backup" compartment for attachments and appendices.

Periodic Progress Reports are written at regular intervals (usually weekly, biweekly, or monthly) and concern projects spanning several months or years.

The writing compartments are the same for both reports, although there are differences in their application. They evolve from the basic report writing pyramid, with two of the compartments relabeled to suit a progress-reporting situation (see Figure 4-2):

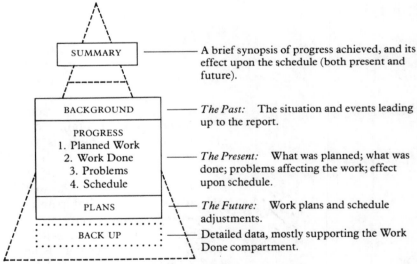

Fig. 4-2. Writing compartments for a progress report.

PROGRESS replaces the basic Facts and Events compartment, and is subdivided into four smaller compartments describing:

- Planned work
- Work done
- Problems encountered
- Adherence to schedule

PLANS replaces the original **Outcome** compartment. There is also an

36

optional BACKUP compartment, for assemblin[...]
pertinent to the project.

How these compartments are used is [...]
explanatory cross-references to the two pro[...]

Occasional Progress Report

Occasional progress reports apply to short projects during which probably only one progress report will be necessary. Sometimes they are written near the midproject point. Occasionally they are written to forewarn management that problems have occurred and delays can be expected. But most often they are written as soon as the project leader has a sufficiently clear picture of work progress to confidently predict a firm project completion date.

The report-writing compartments for an occasional progress report are shown in Figure 4-2 and amplified below.

Summary. The **Summary** should comment briefly on the progress achieved and whether the project is on schedule; it may also predict a project completion date. Its information is drawn from the Work Done, Schedule, and Plans compartments.

Background. If the report will be read only by persons familiar with the project, then only minimum **Background** information is necessary. But if it will also be circulated to other readers, then the Background should briefly describe the persons involved in the project, and the location and dates (i.e., it should answer the questions Who?, Where?, Why?, and When?).

Progress. The **Progress** compartment contains information from the four subcompartments described below, which normally are arranged in the order shown here (it is not uncommon, however, for some of these subcompartments to overlap or be omitted):

1. **Planned Work** outlines what work should have been completed by the reporting date. Normally only a brief statement, it can refer to an attached schedule or work plan.
2. **Work Done** describes what work was actually completed.
   - Brief comments are sufficient for work that has gone smoothly and progressed as planned. If lengthy numerical data are included, they should be attached to rather than inserted in the narrative.
   - Detailed comments should be made on any variances from the plan. They should explain why the variances occurred and any unusual action that was taken.
3. **Problems** are events or situations which affected the *doing* of the job (e.g., a blizzard that stopped work for two days, late delivery of essential parts, or a strike which prevented access to necessary data). They should be described in detail, and should include what action was taken to overcome each problem and what success the action achieved.
4. **Schedule** states whether the project is ahead of, on, or behind schedule. If ahead of or behind schedule, the difference should be quoted in hours, days, or weeks.

Plans. This usually short compartment describes the report writer's plans and expectations for the remainder of the project. It should indicate whether the project will finish on schedule and, if not, predict a revised completion date. There should be an obvious link between this compartment and the previous subcompartment (Schedule).

Backup. The optional **Backup** compartment contains data such as statistics, results of tests, specifications, and drawings, which if included with the earlier parts of the report would tend to clutter the report narrative. This supporting information is grouped and placed in "attachments." Each attachment must be referred to in the Background or Progress section of the report, so that the reader will know it is there.

These compartments are identified in Progress Report No. 1.

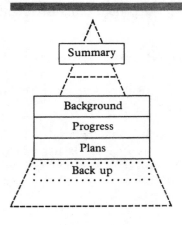

PROGRESS REPORT NO. 1:
Occasional Progress Report

Marjorie Franckel has been delayed in carrying out environmental studies in the Yukon and reports her progress to Vic Braun, her manager. Her progress report was originally handwritten, because she is working on a wildlife study in a remote area and has no access to a typewriter. It has been typed for insertion in these guidelines.

1. This is the **Summary**, and it reports mainly that Marjorie's study is running behind schedule.

2. This **Background** compartment also includes **Planned Work**. In an informal report written under such difficult conditions, Marjorie could have been forgiven if she had omitted this paragraph. She has included it because she knows that Vic Braun (her department manager) probably will have her report typed and sent to the Department of Transport.

3. The **Work Done** compartment starts here. Facts and figures such as these make a report writer sound confident and knowledgeable.

4. The attached map becomes the **Backup** compartment. (It has been omitted from this copy of the report to conserve space.)

5. **Work Done** continues here.

6. In this **Problems** compartment, Marjorie outlines why part of her report is rather indefinite and the study has been delayed.

7. The **Schedule** compartment is this single, rather indefinite sentence. Marjorie cannot be more exact because she simply does not know how long it will take to find and interview people.

8. This final paragraph is the **Plans** compartment.

H. L. Winman and Associates

INTER - OFFICE MEMORANDUM

From: Marjorie Franckel, Biologist   Date: August 18, 19xx

To: Vic Braun, Manager   Subject: Progress: Study of
Environmental Studies   Caribou Calving Grounds

(1) My study of the calving areas used by the Porcupine herd of caribou has been delayed by lack and inaccessibility of data. I doubt whether I will be able to complete the study before September 15.

(2) The study is being done for the Department of Transport, to determine the boundaries and dates of calving so that specific areas can be designated as "Restricted Flying Zones" during the calving season. Currently I am working out of Old Crow in the Yukon.

(3) I have defined the eastern and western limits of the North Slope calving area bordering the Beaufort Sea (see attached map), and have identified three approach routes used by the caribou during their northbound spring migration. These are: (4)

1. Through the Richardson Mountains in the east, along the Yukon/Northwest Territories border.

2. Through the Brooks Range of mountains north of Old Crow.

3. Through the Brooks Range in Alaska, between the Canning River and the Yukon border.

(5) In normal years most calving seems to take place in the Arctic Wildlife Refuge in Alaska between early May and early June. But if bad travel conditions delay the migration, calving occurs farther east along the coastal plain or sometimes even in the mountain ranges while the herd is still migrating.

(6) My problem has been to identify which migration routes are most used, clear-cut dates when calving occurs, and the earliest and latest dates that the caribou have been known to enter the coastal plain. Only a few residents have observed calving, and I have been trying to identify who they are and to interview them. This lack of real information has delayed (7) my study by at least 15 days.

(8) For the next two or three weeks I will be travelling with an interpreter to interview Inuit in very small communities north of Old Crow and as far east as Aklavik. During this period it is unlikely you will be able to contact me.

Marjorie

Periodic Progress Report

The compartments for a Periodic Progress Report contain similar information to those for an Occasional Progress Report, but there is some shift in content and emphasis.

The format of a Periodic Progress Report also appears to be more rigid than that of an occasional report. This rigidity is imposed not so much by established rules as by the content and shape of the initial report in a series. The implication is important: Report writers should take great care in planning a progress report which is to be the first of a series, because they will be expected to conform to the same shape in successive reports.

The compartments outlined below are those shown in Figure 4-2. They provide useful guidelines to follow and demonstrate the differences in content between the occasional and periodic progress reports.

Summary. The **Summary** should comment briefly on the work accomplished during the reporting period. It may also mention whether the project is on schedule. This information can be drawn from the Work Done and Schedule compartments.

Background. Except for the first report in a series, which will be fairly detailed, the **Background** compartment probably will refer only to

- The project number or identification code
- The dates encompassing the specific reporting period
- The situation at the end of the previous reporting period, with particular reference to the project's position relative to the established schedule

Progress. The **Progress** compartment is divided into four subcompartments. In short reports these subcompartments may interlock or overlap, but in longer reports they are more likely to be independent units. If there is no information for a particular compartment, then the compartment is simply omitted.

1. **Planned Work** outlines what should have been accomplished during the reporting period. It may refer to either the original schedule or a revised schedule defined in a previous progress report. It is usually short, sometimes it is combined with Work Done, and occasionally it can be omitted.
2. **Work Done** describes what has been achieved during the reporting period. Ideally, this subcompartment will
   - Open with a brief overview statement which sums up in general terms what has been accomplished.
   - Continue with a series of subparagraphs each describing in more detail what has been done on a specific aspect of the project.
   - Refer to attachments containing comprehensive numerical data, statistics, or tables (see Backup compartment, below).
   - Explain variations from the planned work, or unusual activities affecting work progress (this may be linked with the Problems subcompartment).
3. **Problems** are factors affecting the work, which probably have caused changes in plans or the schedule. The report should describe what action has been taken to overcome the problems, whether the action was successful, if the problems still exist, and what action will continue to be taken, either to avert the problem or to make up lost project time.

4. **Schedule** states whether the project was ahead of, on, or behind schedule on the last day of the reporting period. (There may be a convenient link between this compartment and the end of the previous compartment.) If ahead of or behind schedule, it should state the number of hours, days, or weeks involved. It may also
   - Predict when the project will be back on schedule
   - Recommend a revised schedule for the next reporting period

Plans. This compartment is very short if the project is running smoothly and is on schedule. But if there are problems affecting the work, it should outline the report writer's plans and expectations for the next reporting period, or even suggest a revised schedule for the whole project.

Backup. This optional compartment is used to store detailed information such as forms containing weekly summaries of work done (e.g., yards of concrete poured, number of panels installed, quality of cable strung), tests, and inspections.

Heading and Paragraph Numbering

Periodic progress reports often can benefit from the judicious use of headings and a simple paragraph numbering system. These headings are suggested:

Summary
Introduction
Project Progress
Problems Encountered
Adherence to Schedule
    Current
    Predicted
Attachments

If a paragraph and subparagraph numbering system also is used, it will

1. Show the reader how the information has been organized
2. Provide a handy means for cross-referencing within the report and between successive reports

The suggested numbering system uses whole numbers for paragraphs and decimals for subparagraphs:

3. <u>Project Progress</u>   *(overview)*
    During the month we . . .
    3.1 Field tests were conducted . . .   *(specific*
    3.2 Three pads were installed . . .   *details)*
    3.3 Old cables were removed . . .

The periodic progress report which follows shows how writing compartments, headings, and a paragraph numbering system can be combined to give shape to a report.

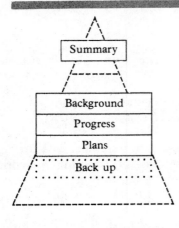

PROGRESS REPORT NO. 2:
A Periodic Progress Report

Comments

1. Project coordinator Roger Korolick has chosen this semiformal format, rather than a memorandum, for his report because it will be circulated to several people, such as the divisional vice-president and the marketing manager.

2. This information takes the place of the **Background** compartment.

3. Roger has taken main points from the *Work Done* compartment for his **Summary**. ("Synopsis" is a lesser-used name for "Summary.")

4. The four headings and the paragraph numbering system help readers *see* how Roger has organized his report.

5. This overview statement at the start of the **Progress** compartment gives the reader a quick picture of the overall situation and introduces the details that follow.

6. Roger places this item first, because it has been carried forward from the previous month (his May 31 report mentioned that the trailer had been taken out for painting).

7. The remaining four items all describe work which continues into the following months. For each, Roger finishes with a closing statement which comments on the schedule, predicts a revised completion date, or states what action will be taken, and when. It is simpler to do this for each item than to lump all the information into one big **Plans** compartment further on in the report.

8. The second of these "unsatisfactory conditions" introduces a problem, which Roger develops further in paragraph 3.1 (on page 2 of the report).

9. Roger must remember to carry this item forward and comment on it in the Progress section of his July report.

10. This is the **Problems** compartment. In paragraph 3.1 Roger forewarns management of a situation which may develop into a serious problem. He must comment on it again in next month's report, as he has done in paragraph 3.2 for the previous month's problem.

    A previous month's problem normally should be mentioned before new problems are introduced, which means that the problem in paragraph 3.2 should be described before the problem in paragraph 3.1. Roger has chosen not to do so because he does not want to interrupt the natural continuity between paragraphs 2.5 and 3.1.

11. This section combines both the **Schedule** and **Plans** compartments. They are combined under one heading because the two topics are interrelated.

    Note how Roger's closing statements in paragraphs 2.3 and 2.4 prepare readers for the changes in plans he announces in paragraph 4.1.

12. The attachment mentioned here forms the **Backup** compartment. (It has been omitted from the report to conserve space.)

VANCOURT BUSINESS SYSTEMS INC

PROGRESS REPORT No. 5 - PROJECT W16

EQUIPPING MOBILE TRAINING AND DISPLAY TRAILER

Reporting Period: June 1 to 30, 19xx

1. Synopsis

    The project is generally on schedule. Trailer painting is complete, air conditioning has been installed, and work has started on wiring and positioning the work stations.

2. Work Accomplished

    With the exception of the instructor's console, work has progressed well during the month. Major activity has occurred in five areas:

    2.1 Exterior Painting. The 24 ft Fruehauf trailer, which had been taken to Display Signs Inc on May 26, was returned on June 8 with the corporate logo and trailer identification painted on both sides and the back.

    2.2 Individual Learning Centers. The first three of the six carrels being built by Wyvern Carpentry House were received on June 15. Our electrician has wired up two of them, and they have been installed on the left-hand wall of the trailer. This component of the project is on schedule.

    2.3 Instructor's Console. Snags have again interrupted construction. Frank Dartmouth, in custom manufacturing, has identified the main problem as late delivery of modules from Capricorn Electronics in San Diego, which has set his assembly schedule back by 15 workdays. He now predicts a completion date of August 19.

    2.4 Display Booths. The two booths for displaying Vancourt equipment are complete and ready for installation in July. Marketing has assembled and tested the display equipment, and I have arranged for the two systems to be installed during the first two weeks of August.

    2.5 Air Conditioning. The air conditioning unit was installed by the Kool-Air Company between June 12 and 16. Initial use of the air conditioner has identified two unsatisfactory conditions:

    a) Moisture is running back into the trailer and dripping onto one of the newly installed carrels.

    b) There is severe vibration and noise from the air conditioner.

    The installation contractor has examined the air conditioner and will return on July 5 to correct these conditions.

2

 3. **Problems Encountered**

    3.1 I am concerned that the air conditioning unit will create a noise problem which will be unacceptable for the type of training we are planning. Already, installation personnel working in the trailer have commented that the noise is unusually high. Although the contractor has assured me that the noise level will drop significantly when the vibration problem is corrected and the carrels, booths, and carpet are installed, I doubt that the drop will be sufficient. We may have to find a means for further lessening the noise.

    3.2 The intermittent short-circuit condition in the trailer strip lighting, described in para 3.3 of my May report, was traced to a faulty switch and corrected on June 6.

 4. **Scheduling**

    4.1 To overcome the problem created by late delivery of the instructor's console, I am planning to advance the installation of the display booths by three weeks (to start on August 1), and to delay installation of the instructor's console until August 22. These changes are shown on the revised work plan attached to this report.

    4.2 I expect the project to be completed by September 15, as scheduled.

R. Korolick

Roger Korolick
Project Coordinator
July 2, 19xx

These guidelines provide only a basic framework for designing periodic progress reports. The framework can be adapted and shaped to suit individual requirements—just as Roger Korolick has adapted it—until the most efficient method is found for reporting progress.

SHORT INVESTIGATION REPORTS

Most investigation reports are longer reports which examine a problem or situation, identify its cause, suggest corrective measures or ways to improve the situation, evaluate the feasibility of each, and select which is most suitable. These are discussed in Chapter 6. There are occasions, however, when only a minor or local problem is examined, and only a short, informal investigation report is needed to describe it. Such reports are described here.

The short Investigation Report has the four basic compartments described in Chapter 2, plus the optional BACKUP compartment. These compartments are illustrated in Figure 4-3 and outlined in more detail below:

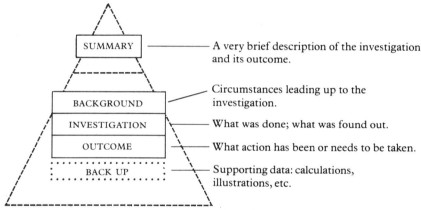

Fig. 4-3. Writing compartments for a short investigation report.

- A **Summary** briefly identifies the problem and how it was, or can be, resolved.
- The **Background** compartment outlines what caused an investigation to be carried out.
- The **Investigation** compartment describes the steps taken to establish the cause of the problem and find a remedy.
- The **Outcome** describes what has been done to resolve the problem or, if other persons have to take the necessary action, recommends what should be done.
- The optional **Backup** compartment stores detailed supporting data evolving from the previous three compartments.

Very short investigation reports are usually issued as interoffice memorandums, or occasionally as a letter. An example follows.

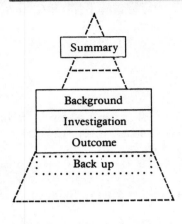

SHORT INVESTIGATION REPORT:
Correcting an Electrical Problem

In this uncluttered one-page report, Tom Westholm places his information confidently into the report writer's pyramid's five compartments.

1. Although the subject line is a little vague, the word *correcting* serves a useful purpose because it implies that Tom has found a solution to the problem.

2. Tom's **Summary** identifies the problem, states its cause, reports that it has been resolved, and suggests what else should be done.

3. The **Background** compartment describes events leading up to the investigation.

4. In his **Investigation** compartment, Tom describes his approach to the problem and what he has discovered.

5. The **Outcome** describes how he corrected the problem, but suggests a better alternative.

6. Here, Tom refers to his **Backup** information (which, to conserve space, has been omitted).

7. This final question is part of the **Outcome**.

Compare Tom Westholm's short informal investigation report with Tod Phillips' five-page semiformal report in Chapter 6. Note how the five compartments used for the short report are expanded to develop more information for the longer report.

INTER-OFFICE MEMORANDUM

To: C. Meaghan, Plant Manager   Date: July 9, 19xx

From: Tom Westholm   Ref: Correcting Electrical
      Maintenance Electrician   Blackouts

(1)

(2) I have traced the recent electrical power failures to a wiring error which created a power overload. Although I have corrected the problem, a better solution would be to install a separate power panel for two of the air conditioners.

(3) The failures started after the air conditioners were overhauled in May, and even then they occurred only infrequently and at random intervals. On every occasion simply resetting the circuit breakers corrected the failure, which made the cause difficult to identify.

(4) As I suspected the air conditioners, I compared the wiring connections against the manufacturer's wiring diagrams but could find no fault. I then examined the four air conditioners in turn, and identified a disconnected load splitter behind air conditioner No. 2. The load splitter was installed six years ago, to prevent the circuit from being overloaded should more than two air conditioner compressors cut in at the same time. Apparently the overhaul contractor failed to reintroduce it into the circuit when re-installing the air conditioners in May.

(5) I have reconnected the load splitter, but suggest we could obtain better performance from the air conditioners if we were to install a new power control box and connect two of the air conditioners to it. We could then remove the load splitter. The cost would
(6) be $645.00, as detailed on the attached cost estimate.

(7) May I have your approval to buy the necessary parts and do the installation?

Tom.

III
SEMIFORMAL REPORTS AND PROPOSALS

5
Test and Laboratory Reports

Considerable variation exists in the presentation of test and laboratory reports (often called *lab reports*). Some laboratory reports simply describe the tests performed and the results obtained, and comment briefly on what the results mean. Others are much more comprehensive: they open with a synopsis of the tests and results; they continue by presenting full details of the background, purpose, equipment, methods, and results; and they finish with an analysis from which their authors draw a conclusion. The more comprehensive laboratory report is described here, because the shorter, simpler form can be adapted from it.

A third form of laboratory report is used in universities and colleges, where students are asked to perform tests and then write a lab report to describe their findings. Comments on laboratory reports written in an academic environment are included at the end of this chapter.

INDUSTRIAL LABORATORY REPORTS

Industrial laboratory reports are based on the report writer's pyramid described in Chapter 2, but the FACTS compartment is expanded to encompass four subcompartments: **Equipment and Materials**; **Test Method**; **Test Results**; and **Analysis**. There is also a **Backup** compartment for holding specifications, procedures, and details of test measurements the author refers to. The major compartments and subcompartments are illustrated in Figure 5-1 and are described in more detail on the pages that follow.

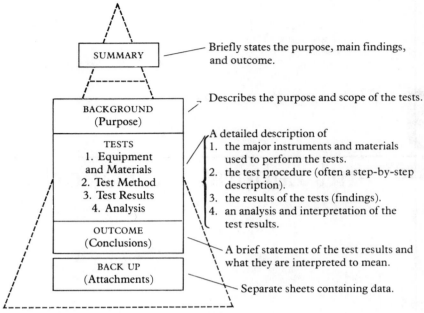

Fig. 5-1. Writing compartments for a test or laboratory report.

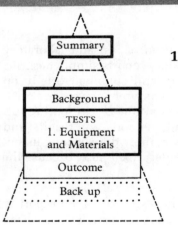

INDUSTRIAL LABORATORY REPORT:
Testing a Water Stage Manometer (Pp. 55–59)

1. Some organizations use a prepared form for the first page of their laboratory reports. The form has spaces for entering predetermined information, such as:

    Title and purpose of test
    Name of client
    Authority for test (i.e., purchase order, letter, etc.)
    Summary of test results
    Signature and typed name of person performing test
    Signature of manager or supervisor approving test results
    Date tests were completed

    Report authors using a form find their **Background** compartment becomes much shorter, or even nonexistent, because many background details are entered in the prepared spaces and need not be repeated.

2. The **Summary** establishes what test was undertaken, sometimes why it was necessary, the main finding(s), and the result.

3. The **Background** compartment starts here. Carole Winterton (the laboratory technician who performed the tests and wrote the report) has chosen to divide the compartment into two parts, each with its own heading: *History* and *Purpose of Test*.

53 *Test and Laboratory Reports*

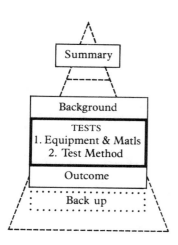

4. This is the start of both the large **Tests** compartment and the *Equipment and Materials* subcompartment.

   The amount of information provided in the *Equipment and Materials* compartment depends on several factors. If the client may want to duplicate the tests or know more about how they were undertaken, or if the person performing the tests needs to demonstrate the extent of testing, a full description is provided; but if the client is more interested in results, and not likely to be concerned with how the tests were run, then only essential details are included. (The same guideline applies to the *Method* compartment.)

   If the equipment setup and list of materials are complex or lengthy, they can be placed in an attachment.

5. A simple diagram showing how the test equipment is connected can help a reader visualize the test setup. If the illustration of a test setup is too large to fit on a standard page, it should be placed at the back of the report, labeled as an attachment, and referred to at the appropriate places within the report.

   If several tests are performed, each with a different arrangement of test equipment, it is better to insert a series of diagrams in the report, each positioned immediately ahead of or beside the appropriate test description.

6. The *Method* compartment describes how the tests were carried out. It can range from a brief outline of the test method used (for a nontechnical reader interested primarily in results) to a step-by-step description of the procedure (for a reader who wants to know how comprehensive the test was). Carole has used a fairly detailed step-by-step description for her report, because her analysis of the results will depend on the reader's full understanding of what was done during the tests.

7. This initial paragraph is a section *summary statement* preceding a fairly lengthy section of the report. Preparing readers to expect a certain arrangement of information helps them more readily accept the facts a report writer presents. By numbering and naming the three tests, Carole is silently saying: "These are the tests you will read about next, and this is the sequence in which I will be presenting them to you." But now she must take care to describe them in the same sequence; if she does not, her readers would likely become disoriented as soon as they encounter the unexpected arrangement.

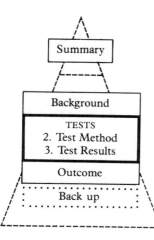

8. Each test is numbered and given a subheading similar to that used in the section summary statement (7).

9. When a process or procedure is lengthy, or is not essential to a full understanding of the report, the steps may be printed in an attachment and simply referred to in the report. (Also see comment No. 16.)

10. Carole has chosen to place these five steps within the report, rather than in an attachment, because she wants readers to be fully aware that steps (a) and (e) were performed *before* they read the test results and her analysis. They support her contention that the fault is unidentifiable and lies within the instrument she is testing.

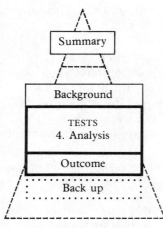

11. The *Method* compartment of a report must be written in clear, direct language which is entirely objective (unbiased, without opinions).

12. The pumping in and out of water has a direct bearing on, and is referenced in, the *Results* compartment. Consequently it must be described thoroughly here.

13. The *Results* compartment describes the major finding evolving from the tests. Like the tests described in the *Method* compartment, the test results must be written objectively.

14. In the **Analysis** compartment a report writer is expected to examine and interpret the test results, and to comment on their implications. The analysis should discuss various aspects influencing or evolving from the tests, and show how they lead to either a logical conclusion or an unanticipated outcome. This helps readers to understand and more readily accept the conclusions which follow.

15. This is the **Outcome** compartment. It should answer the question, resolve the problem, or respond to the request identified in the **Background** compartment (titled *Purpose of Test* in Carole's report). It must never introduce new data or present information which will surprise the reader.

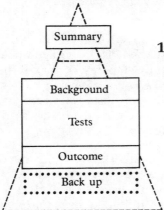

16. Attachments form the **Backup** compartment. Their purpose is to provide a place for storing data which a reader does not need while reading the report but may want to inspect later. They may comprise a detailed procedure used during the tests, a lengthy table of test results containing measurements and dial readings, or photographs, sketches, and drawings.

    All attachments must be referred to in the report (the attachment on the facing page is referred to near the foot of page 2 of the report). They should be presented in the sequence in which they are mentioned in the report, and then numbered sequentially as "Attachment 1," "Attachment 2," etc.

    NOTE: To conserve space in these guidelines, only Attachment 1 is included with this report.

ENVIRONMENTAL TEST LABS INC

Test Report No. 34/07
June 14, 19xx

Summary of Test Results

Pressure tests of a Caledonia Water Stage Manometer model WSM, serial No. 2306, show that although all components are operating satisfactorily the manometer apparently has a minute, unidentifiable gas leak.

History

The manometer was shipped to the Test Lab from site 24 of the Agassiz Water Control System (AWCS). It was removed from service on May 22, following a visiting technician's report that the manometer was recording erratic, sharp changes in water stage which contradicted his visual observations of water levels. No tests of the manometer, or of the gas delivery system between it and the underwater orifice, could be performed in situ because of the remoteness of the site.

Purpose of Test

In AWCS memorandum 0693 dated June 5, 19xx, we were requested to test the manometer, determine whether the manometer or the site's buried gas delivery system is causing the problem, and, if the manometer proves faulty, identify the cause.

Equipment Set-up

The manometer was installed on a workbench, levelled $18°$ from the vertical, and connected to a mock-up of the site's gas delivery system, which consisted of:

a) A cylinder of super-dry nitrogen, through a gas flow regulator.

b) 108 ft of 1/8 in. ID polyethylene tubing terminating in a water tank, with the tube's orifice submerged 38 in. below the surface (to simulate site conditions).

c) A Franck and Corwin type B-37 continuous recorder.

The manufacturer's operating manual, publication 6425, was used as a reference. The test hook-up is shown in figure 1.

1

⑤

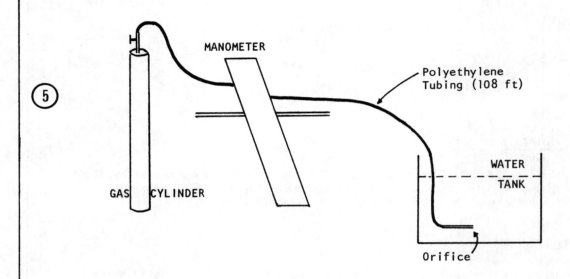

Fig. 1. Water Stage Manometer Test System

⑥ Method

⑦ Tests were made of 1) the nitrogen gas delivery system, 2) the manometer, and 3) the total system.

⑧ 1. Tests of Nitrogen Gas Delivery System

The gas flow regulator of the nitrogen cylinder was first connected directly to the polyethylene tubing, the orifice in the water tank was temporarily plugged, and gas was allowed to flow into the system until the regulator gauge read 70 psi. The tubing was then inspected for leaks; none were found and the pressure remained stable for the ensuing two hours of the test.

⑨ The manometer was reconnected to the system and nitrogen gas was applied as specified in step 8 of the manufacturer's manual (see attachment 1). When a flow rate of 15 bubbles per minute had been achieved in the sight feed, the flow rate was checked at the orifice. It was a steady 8 bubbles per minute, which conforms to the manufacturer's specification of one-half the flow rate at the sight feed, plus or minus 15%.

The sight feed and orifice bubble rates were checked at two-hour intervals during the tests and remained within specification throughout.

2

2. <u>Check of Water Stage Manometer</u>

All external parts of the manometer were checked for proper operation:

a) The bubble flow rate in the sight feed was observed at two-hour intervals during the system checks, and at 15-minute intervals during the final four hours. It remained constant and within specifications.

b) The float switch contacts were examined and found to be clean.

c) The set screw on the servo control was adjusted until the drive motor moved away from the correct setting. The servo control operated smoothly to follow the adjustment.

d) The constant-differential pressure regulator was removed and cleaned. No foreign matter was found.

e) A soapy solution was applied to all exposed connections, both on the manometer and to the gas delivery system. No bubbling occurred.

3. <u>Check of Total System</u>

The system was allowed to stabilize for 3 hours, and then run continuously for 30 hours. For the first 26 hours the water in the tank was pumped in and out at controlled rates, to simulate changes in water stage:

a) Water was pumped out at 2.75 gallons per minute (gpm). After 8 hours the head of water above the orifice had decreased from 38 in. to 16 in.

b) Water was then pumped in at 3.0 gpm for 12 hours, after which the water head had increased to 52 in.

c) Water was again pumped out, this time at 2.33 gpm for 6 hours, until the water head had returned to 38 in. above the orifice.

The system was run for a further 4 hours, with no change in water level, during which further manometer checks were performed. Throughout, the manometer appeared to operate correctly.

<u>Test Results</u>

Tests of the gas delivery system and of the water stage manometer showed no apparent faults. But when the chart on the Franck and Corwin B-37 recorder was removed and inspected at the end of the tests, it showed that a fault existed somewhere in the system. Instead of recording a steady decrease, increase, and then decrease of water head, the trace on the chart displayed a series of "steps," indicated by apparent abrupt decreases of water level, each followed by a slow recovery (see attachment 2). These "false troughs" were present for increasing, decreasing, and stable water level conditions.

3

Analysis of Results

(14) False troughs are caused by minute intermittent leaks in the total gas purge system, resulting in a temporary loss of pressure. They appear on the chart as a comparatively rapid drop in water level followed by a slow recovery, usually of about one hour. Intermittent leaks are more likely to occur at high water stages, with the result that crests are recorded several feet below their true stage, although this was not apparent in our tests. Very small intermittent leaks can be extremely difficult to locate.

We believe the leak is within the manometer, rather than in the gas delivery system. The pre-test pressure check of the gas delivery system, and the soap test of its connections, produced no evidence of leaks between the manometer and the underwater orifice.

The erratic water level readings reported by the site and the false troughs identified during the tests are probably different interpretations of an identical fault. Since a different gas delivery system was used in each case, the problem is more likely to be within the manometer.

Conclusions

(15) Our tests show that the erratic water stage readings recorded at AWCS site 24 were probably caused by a tiny, undetectable internal gas leak in the water stage manometer. The site's gas delivery system is less likely to have caused the fault.

Tests performed by:

Carole Winterton
Carole Winterton
Lab Technician III

Approved by:

F L Cairns
Frederick L. Cairns, P.E.
Supervisor, Tests & Procedures

Attachment 1

EXTRACT FROM MANUFACTURER'S OPERATING MANUAL
FOR WATER STAGE MANOMETER   MODEL WSM

8. Instructions for Purging the System

To purge (introduce nitrogen gas into) the system, proceed exactly in the following sequence:

8.1 Check that the following valves are closed:

Valve	Rotation
Feed pressure adjustment screw	fully CCW
Feed rate adjustment needle valve	fully CW
Manometer shut-off valve	fully CW

8.2 Turn the bubble tube shut-off valve fully CW (open).

8.3 Turn the nitrogen cylinder valve fully CCW (open).

8.4 Rotate the feed pressure adjustment screw slowly CW until the pressure gauge reads 35 psi.

8.5 Rotate the feed rate adjustment valve slowly CCW until bubbles can be seen flowing in the manometer sight feed. Adjust the valve until the flow rate is 15 bubbles per minute.

8.6 Check that the flow rate at the orifice is 7 or 8 bubbles per minute.

5

ACADEMIC LABORATORY REPORTS

Laboratory reports written in an academic setting use the same writing compartments as those written in industry (see Figure 5-1), but there is a shift in purpose and emphasis. An industrial laboratory report responds to a specific request or demand, and so answers a question or meets a stated need. An academic laboratory report more often is used as a vehicle for helping students learn how to perform a particular test, understand a process or procedure, prove a hypothesis, or test a theory. Hence it usually does not respond to a demand (other than a professor's request) or meet a specific need. It may, however, answer a question.

Each university or college's science or engineering department has its own requirements for lab reports, which makes it difficult to specify exact writing compartments here. Those described below offer the most generally accepted approach.

Summary. A brief statement of the purpose or objective of the tests, the major findings, and what was deduced from them.

Objective or Purpose. A more detailed statement of purpose or objective, plus other pertinent background data. (This writing compartment may be combined with the **Summary** if there is little background information.)

Equipment Setup. A list of the equipment and materials used for the tests, and a description and illustration of how the equipment is interconnected. (If there is a series of tests requiring different equipment configurations, a full list of equipment and materials should appear here. A description and illustration of each setup should then be inserted at the beginning of each test description.)

Method. A step-by-step detailed description of each test, similar to the **Test Method** section of an industrial laboratory report. Attachments may be used for lengthy procedures or process information.

Results. A statement of the test results or findings evolving from the tests.

Analysis. A detailed analysis of the results or findings, their implications, and what can be learned or interpreted from them.

Conclusions. A brief statement describing how the tests, findings, and resulting analysis have met the objective stated in the **Objective or Purpose** compartment.

Data (Attachments). A separate sheet (or separate sheets) containing data derived during the tests, such as detailed calculations, measurements, weights, stresses, and sound levels. Lengthy procedures or process descriptions are sometimes included as attachments.

If several tests are performed, and there are results from each, it may be better to have separate **Equipment Setup**, **Method**, and **Results** compartments for each test. The organization plan would then look like this:

Summary
Objective or Purpose
Equipment and Materials
Test No. 1:
    Equipment Setup
    Method
    Results
Test No. 2:
    Equipment Setup
    Method
    Results
Test No. 3:
    Equipment Setup
    Method
    Results
Analysis
Conclusions
Data (Attachments)

6
Investigation and Evaluation Reports

An investigation report describes a problem or situation that has been investigated, examines methods for correcting the problem or improving the situation, and usually suggests what should be done about it. If the report evaluates alternatives (for example, an examination of several sites for locating a fast-food restaurant), it may be called an evaluation report. Or if it examines the practicality of doing something new or different, it may be called a feasibility study, as discussed in Chapter 8. And sometimes it is wrongly called a recommendation report, a misnomer which should not be used as a label for a particular type of report, because almost any report can make a recommendation.

Investigation reports can be as short as only one or two pages, but more often they are much longer, ranging from three to 30 or 40 pages. Shorter reports are normally typed as interoffice memorandums or letters, while longer reports may have a title centered at the head of the first page (which makes them eligible to be labeled semiformal reports). Some are even given the full formal report treatment shown in Chapter 8.

A very short investigation report has five compartments, which are shown in Figure 4-3. For longer investigation reports the compartments are expanded to introduce subcompartments as shown in Figure 6-1. These compartments and subcompartments are identified and described in greater depth in the comments for the five-page investigation report which follows.

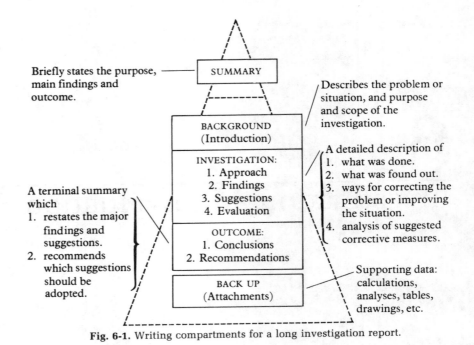

Fig. 6-1. Writing compartments for a long investigation report.

SEMIFORMAL INVESTIGATION REPORT:
Study of High Gas Consumption (Pp. 69–73)

Throughout his report author Tod Phillips keeps his readers firmly in mind. He is aware that, although he has done his investigation and is preparing his report for the Marsland Construction Company, his real readers are the homeowners, Mr. and Ms. Parsenon. It is for them that he so carefully develops his case, and uses language and terminology which they will readily understand.

If Marsland Construction Company had simply asked Tod to investigate the cause of high gas consumption, and if he knew that only they would use the information, then he could be much more direct. He could tell them in one sentence that he has checked the gas furnace, hot air ducts, gas flow meter, and insulation, and found them all to be satisfactory. And then he could go straight into his comparative analysis of gas consumption records and thermostat settings.

But Tod has been briefed by Marsland Construction Company that the report is needed specifically for Mr. and Ms. Parsenon, who have complained of high gas consumption ever since the house was new and they first occupied it. Over the years Marsland has done numerous checks, made many adjustments, and found little wrong with the house. And yet the Parsenons have still complained. Finally, Marsland called in a professional consultant and asked him to carry out an independent study which would identify the cause of the problem and recommend how it could be resolved. Marsland could then present the results to Mr. and Ms. Parsenon, to persuade them that the construction company had done all it can to rectify the problem. As Jack Marsland told Tod Phillips during their initial discussion,

> Whatever you discover is wrong, we'll fix it if it's our fault. Then I can use your report to get the Parsenons off my back!

Tod Phillips has written a semiformal report, although its length is such that he could also have prepared it in letter form. Some comments on the implications of using either format, and of addressing the report directly to the homeowners rather than to the construction company, follow the report.

The comments which start below refer to the corresponding numbers marked on page 1 of the investigation report.

1. A title or main heading should be noticeable and informative: it should describe what the report is about. Too often, a report title can leave readers wondering what the report covers; for example, "Gas Consumption Investigation Report" would have been an inadequate title in this situation.

2. A **Summary** carries the report's highlights, stated as briefly as possible. For an investigation report the Summary should cover:

    a. the purpose of the investigation ("*Our investigation of heating fuel consumption . . .*"),
    b. the major findings ("*. . . consumption . . . only slightly higher than in comparable homes . . .*"), and
    c. the major conclusions and/or recommendations ("*. . . consumption can be reduced . . . by maintaining a lower home temperature . . . and insulating . . .*").

    In a long report like this, the Summary often is written *after* the rest of the report. The report writer can then extract the highlights from the **Background**, **Facts and Events**, and **Outcome** compartments.

3. When headings are used in a report, the **Background** compartment is titled "Introduction." An Introduction usually contains three pieces of information, which need not appear in this order:

    - Events leading up to the investigation.
    - The purpose of the investigation.
    - The scope of the investigation.

4. The **Investigation** compartment starts here by outlining the **Approach**. It explains that the study had two phases, which prepares the reader to find these two phases treated separately. Indeed the heading immediately following this paragraph tells the reader that phase one is about to start.

5. Not all of the headings exactly parallel the compartments used to write the report. To place a heading at the beginning of each compartment would have made the report too rigidly structured. Tod Phillips used the compartments to ensure that he was organizing his report properly, and then later inserted headings where they would help readers see the logical divisions of information. Headings which most often parallel the report-writing compartments are:

    Summary (Main Message)
    Introduction (Background)
    Conclusions and Recommendations (Outcome)
    Attachments (Backup)

6. This short paragraph continues the **Approach** compartment by identifying what physical checks will be carried out in the Parsenon home. (Later in the report a second **Approach** paragraph will describe how the comparative analysis was carried out.) An opening overview paragraph like this is useful because it prepares the reader to expect a certain narrative sequence.

7. The **Investigation Findings** compartment starts here and continues through four paragraphs. (A second set of findings appears later in the report.)

8. Tod has found that the problem lies entirely with the homeowners, and not with the construction company. So he has to tell Mr. and Ms. Parsenon something they don't really want to hear: that the builder is blameless and that they have to take whatever corrective action is necessary. Consequently he has to *persuade* them to accept his findings.

   His careful accumulation of *evidence* in this first set of **Investigation Findings** does exactly that. He cannot dismiss the insulation and heating equipment in just a few words, because the Parsenons most likely suspect that both are the cause of their problem. So he carefully examines each aspect, quotes *facts* to prove there are no faults, and in no case allows his personal opinions to intrude. By the end of this section there must be no doubt in the Parsenons' minds that the fault lies elsewhere.

9. This is the second part of the **Investigation Approach**.

   In many investigation reports it is possible to have a straight-through discussion, in which the whole approach is presented first, and is followed by all the findings. Tod has chosen to use a two-stage method because he has two distinct aspects to deal with:

   a. The physical check of the home.
   b. The comparative analysis of gas consumption records.

10. A reader should be able to read a whole report right through without having to turn to the attachments, but should always be informed at appropriate places that supportive evidence is attached, and where to find it.

11. The second set of **Investigation Findings** starts here. The results of this part of the investigation are based on evidence supplied in an attached comparative analysis. Tod presents his findings here without discussion; in effect he is saying to readers: "This is what I found out, and you can verify all of it by referring to the comparative analysis." His use of brief subparagraphs (point form) helps maintain a detached, almost clinical presentation of these facts.

12. In the **Suggestions** compartment a report writer can offer a single suggestion, several suggestions, or alternative suggestions. Tod presents two main suggestions, and implies that one should be adopted and the other is optional.

13. The **Evaluation** compartment allows a report writer to shed some of his or her impartiality. Readers like to form their own opinions of the validity of each suggestion, but need a convincing, logical, rational evaluation on

which to base them. They depend on the report writer to evaluate factors such as

- Feasibility
- Suitability
- Simplicity
- Effectiveness
- Cost

(And sometimes a report writer may also have to evaluate the effect of taking *no* action; i.e., adopting none of the suggestions.)

Tod Phillips has to convince his secondary readers (the homeowners, Mr. and Ms. Parsenon) that they must take at least one step, and preferably two, to resolve their problem. He tries to persuade them to accept his suggestions by

- Mustering evidence to show they have to lower the temperature in their bungalow. (He quotes federal specifications to demonstrate what gas savings they should be able to achieve, and then reinforces his case by referring to local residents who are achieving such savings.)
- Suggesting they would be wise to insulate at least part of their basement. (He acknowledges their problem of cold floors, and then points to what someone else has done.)
- Explaining how they could gain additional savings by installing more insulation in the walls and ceiling of their home. (He recognizes that this would be costly, and so presents it only as an optional extra.)

It would be difficult to refute the validity of Tod's evaluation. It is carefully developed, so that it carries readers logically from one point to the next. By now, Tod's readers should feel they *know* what the conclusions and recommendations are going to be.

14. Tod has written a straightforward evaluation because he is offering several suggestions which will have a cumulative effect. His readers have to decide how many of his suggestions should be adopted, rather than choose among them. He might have arranged his information differently if there had been alternative suggestions.

For example, suppose that you have investigated several word processing systems to identify which would be most suitable to install in your office. Now you have to write an investigation report to management, in which you will recommend the best system. There are two methods you can use for arranging the **Suggestions** and **Evaluation** compartments:

A. Either you can present all the suggestions first and then evaluate them (as Tod has done), like this:

    Suggestions:
        System A
        System B
        System C
    Evaluation:
        System A
        System B
        System C

68 *Investigation and Evaluation Reports*

B. Or you can present each suggestion in turn and then immediately evaluate it, like this:

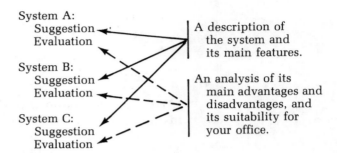

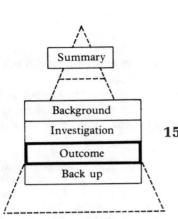

15. The Conclusions and Recommendations form the report's **Outcome** compartment. They present the results of the investigation, suggest what needs to be done, and sometimes recommend what action the reader should take.

    The Conclusions should repeat the main features of the **Investigation Findings** compartment. They must never introduce any information or ideas which have not been discussed previously in the report.

16. Recommendations must be strong and definite. They should be stated in the active voice and first person, rather than the bland passive voice. That is, they should say "I recommend . . ." or "We recommend . . .," and *not* "It is recommended that. . . ." And they should never recommend action which has not already been discussed in the report.

17. The signature at the end of the report is optional. But the date the report is issued should be present, either here or immediately after the report title.

18. Attachments (sometimes called Appendices) hold factual evidence which supports statements made in the report but is too comprehensive, complex, or detailed to be included with the main narrative. They may be calculations, analyses, cost estimates, drawings, photographs, plans, or copies of other documents. They form the **Backup** compartment of the report writer's pyramid. Certain "rules" apply to attachments:

    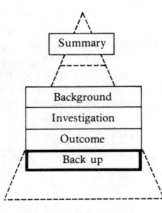

    a. Readers should not have to turn back to them as they read the report (although they may choose to refer to them later). This means that the most significant features of an attachment must be quoted in the investigation narrative.
    b. Every attachment must be referred to in the report narrative. It must serve a useful purpose, and never be included simply because it contains possibly useful information.
    c. The attachments should appear in the order in which they are referred to in the narrative. If there is more than one attachment, they should be numbered Attachment 1, Attachment 2, etc. (or, if they are referred to as appendices, as Appendix A, B, C, etc.).

H. L. Winman and Associates

PROFESSIONAL CONSULTING ENGINEERS
475 Reston Avenue-Cleveland, Ohio, 44104

① STUDY OF REPORTED HIGH GAS CONSUMPTION AT 1404 GREGORY AVENUE

Summary

② Our investigation of heating fuel consumption in Mr. and Ms. R. M. Parsenon's bungalow at 1404 Gregory Avenue shows consumption to be 4.37% higher than in homes of comparable size and construction. We believe consumption can be reduced to a normal or even lower level by maintaining the home at a slightly lower temperature and by insulating the upper four feet of the basement walls. Even greater fuel savings could be achieved by installing additional insulation in the ceiling and walls.

Introduction

③ The investigation followed lengthy correspondence between Marsland Construction Company, who built the bungalow, and the homeowners, who have continually reported excessively high fuel consumption and uncomfortably cold floors. In a letter dated February 16, 19xx, Marsland Construction Company authorized us to carry out an independent study which would determine the extent of high fuel consumption, identify the cause, and suggest remedies.

④ We divided the study into two phases: a physical check of the building and its heating system, and a comparison of gas consumption in the Parsenon home with that of similar bungalows.

⑤ Checks of Building and Heating System

⑥ Our examination of the Parsenon residence involved checks of the gas furnace and hot air ducts, the gas company's flow meter, and the home's insulation.

⑦ The gas furnace and hot air ducts were examined by Montrose Heating and Supply Specialists Inc. Both were found to be in good condition, with the exception of the humidifier plates in the furnace, which were badly corroded. Their poor condition might cause slightly lower humidity in the home but would not affect the temperature. Low humidity, however, might create the impression of a lower-than-desirable temperature. New humidifier plates were installed, and a spare set was left with the homeowners for installation in one year's time.

1

We asked Montrose Gas Company to check the gas flow meter. They reported that it had been replaced twice in the past twelve months at the householders' request, and on both occasions no fault could be found. Neither could they find fault with the present flow meter.

We checked the bungalow's insulation and found:

⑧
a) The ceiling has a minimum of 7 inches of wood chips, which is equivalent to an R16 insulation factor.

b) The walls have Fiberglas insulation with an R8 insulation factor.

c) The basement is unfinished and uninsulated.

Although this level of insulation would not meet government standards for a dwelling built today (i.e. R30 in the ceiling and R18 in the walls), it fully meets the insulation requirements which were in effect when the home was built in 19xx.

⑨ Comparison with Comparable Homes

As our checks of the Parsenon dwelling showed no significant cause for the reported high gas consumption, we decided to compare consumption in Mr. and Ms. Parsenon's home with that of two groups of bungalows of comparable age and size. They comprised:

    Four identical homes built by the same contractor. They were Gregory Avenue numbers 1396, 1399, 1407, and 1410.

    Four homes built at the same time, but by other contractors. They were Gregory Avenue numbers 1506, 1515, 1524, and 1581.

In each case we obtained permission from the homeowners to quote their consumption records for the past calendar year (19xx). We also asked homeowners to inform us if any additional insulation had been installed since their bungalows were built, and the setting at which they kept their
⑩ thermostats. The results are shown in the attached table.

Examination of the comparison table shows that:

⑪
1. Gas consumption in the Parsenon home for 19xx was 38 MCF (38,000 cubic feet), or 4.37%, greater than the average of the eight other homes we evaluated.

2. The thermostat in the Parsenon home was set 2.5°F higher by day, and 7.1°F at night, than the average setting for the eight other homes.

3. Gas consumption for the six homes in which the thermostat setting was lowered at night was consistently lower than the consumption in the three homes in which the thermostat setting was not lowered.

4. Gas consumption in Marsland-built homes was comparable to that in homes built by other contractors.

5. Two homes in which some additional insulation has been installed consumed significantly less gas than homes which have only their original insulation.

Methods for Reducing Gas Consumption

⑫ We believe Mr. and Ms. Parsenon can reduce gas consumption in their home to an average, or even slightly better than average, level at little or no cost. This reduction can be achieved by lowering the thermostat setting from its present constant 73°F, to 70° during the day and 65° at night. They could achieve a further significant reduction, but at extra cost, by installing additional insulation in the ceiling, walls, and basement.

⑬ Although it is difficult to predict an exact saving in gas consumption, Federal Insulation Specification FIS-2820/78 suggests that for every 1000 cubic feet of floor space, each one-degree reduction in house temperature conserves about 10.3 MCF of gas annually (MCF = 1000 cubic feet). In Mr. and Ms. Parsenon's residence, a 3°F reduction in temperature should therefore cut consumption by 30.9 MCF annually. We also believe that a further 8 to 10 MCF of gas could be saved annually by reducing the temperature to 65°F at night. These estimates seem to be borne out by the consumption figures for the homes at 1407, 1506, and 1581 Gregory Avenue, in which lower temperatures are maintained without additional insulation.

Ms. Parsenon has commented that they are forced to keep their home temperature at 73° because at any lower temperature the floor is unbearably cold. We asked the owners of an identical bungalow at 1410 Gregory Avenue, who maintain a 69° temperature in their home, if their floors are similarly cold. Their experience offers a probable solution. After the basement of their bungalow was insulated, they found the floors to be considerably warmer and that they could reduce the temperature from its previous 72° to 69° without discomfort.

⑭ The basement at 1410 Gregory Avenue has been fully insulated. We believe it would be necessary to insulate only the upper 4 feet of a basement to obtain a similar result, because below the 4 ft level the walls are insulated naturally by the surrounding soil. Basement insulation is easily applied by nailing or gluing styrofoam panels directly onto the walls. We estimate that to insulate the upper 4 feet of the basement in the Parsenon residence would cost $450 if the work is done by a contractor, or $180 if it is done by the homeowners.

Installing additional insulation in the ceiling and walls of the Parsenon home would achieve even greater savings in gas consumption, and also enhance the comfort level at the suggested thermostat settings. In 19xx, the fully insulated home at 1515 Gregory Avenue consumed 55.1 MCF less gas than the average for the eight homes we evaluated, and 93.1 MCF less gas than Mr. and Ms. Parsenon's bungalow. Significantly, it is also the largest bungalow in the group.

Because the cost of insulating the ceiling and walls, and even the basement, of their home will depend on the quality (R factor) of insulation they desire, we suggest that if Mr. and Ms. Parsenon want to install additional insulation they obtain the advice of, and an estimate from, a recognized insulation contractor.

Conclusions

Our study shows that gas consumption in Mr. and Ms. Parsenon's home is 4.37% higher than average, and that this high consumption is caused primarily by the dwelling being maintained at higher than average room temperatures.

Consumption could be reduced to average for this type and size of dwelling by lowering the thermostat level $3°$ by day, and $5°$ at night, and by maintaining an optimum humidity level. It could be reduced to better than average by also installing additional insulation.

Recommendations

We recommend that Mr. and Ms. Parsenon reduce the temperature in their home to $70°F$ during the day, and $65°F$ at night, increase the floor temperature by insulating the upper 4 feet of the basement walls, and maintain constant humidity by replacing the humidifier plates in the gas furnace at least once a year.

If this does not achieve the desired reduction in gas consumption, then we suggest that Mr. and Ms. Parsenon install additional insulation in the walls and ceiling of their home.

J E Phillips
Tod E. Phillips
March 16, 19xx

ATTACHMENT

COMPARISON OF GAS CONSUMPTION
IN NINE GREGORY AVENUE BUNGALOWS
FOR CALENDAR YEAR 19XX

House No.	Year Built	Size (sq ft)	Consumption* (in MCF)	Thermostat Setting (Day °F)	Thermostat Setting (Night °F)	Additional Insulation
R.M. Parsenon Residence						
1404	1975	1004	906.4	73	73	None
Identical Homes Built by Marsland Construction Company						
1396	1975	1004	870.6	72	64	None
1399	1974	1004	894.4	70	70	None
1407	1974	1004	880.9	70	63	None
1410	1975	1004	841.3	69	69	Basement
Group Average:		1004	871.8	70.2	66.5	
Nonidentical Homes Built by Other Contractors						
1506	1973	966	868.9	71	62	None
1515	1976	1080	813.3	70	62	Ceiling, Walls Basement
1524	1973	980	900.3	72	72	None
1581	1973	966	877.5	70	65	None
Group Average:		998	865.0	70.7	65.2	
AVERAGE OF 8 CONTROL HOMES		1001	868.4	70.5	65.9	

* MCF = 1000 cubic feet

5

COMPARISON OF SEMIFORMAL AND LETTER FORM INVESTIGATION REPORTS

A semiformal investigation report is generally longer, seems to have more dignity, and appears to be more formal than a letter. Its title is centered at the top, its contents are clearly divided into compartments each preceded by a heading, and its language is usually a little less personal. It can be a useful way to convey information when the contents of a report are meant to influence a third party (in the example, the report is written for Marsland Construction Company but its results are really intended for the homeowners).

Because it is not addressed to a particular reader, a semiformal investigation report often needs to be accompanied by a cover letter. This letter serves two purposes: 1) it very briefly summarizes the results presented in the report; and 2) it provides a place for report writers to make comments they might prefer not to insert in the report. The cover letter that accompanied Tod Phillips' report for the Marsland Construction Company does both:

Dear Mr. Marsland:

Our investigation at 1404 Gregory Avenue shows that Mr. and Ms. Parsenon's complaint of high gas consumption is partly justified. We find, however, that it is caused by the high temperature at which they keep their home rather than by faulty heating equipment or inadequate insulation.

We hope the attached report will help you convince Mr. and Ms. Parsenon to take the appropriate steps necessary to bring gas consumption down to an acceptable level.

Sincerely,

Tod Phillips

A letter-form investigation report, however, does not normally exceed three or four pages, may still contain headings, and addresses the recipient personally. If Tod Phillips' report had been written as a letter to the Marsland Construction Company, paragraph one (the Summary) probably would not have changed, but personal words such as "you," "your," "we," and "our" would have been sprinkled more liberally throughout paragraph two and many of the subsequent paragraphs. For example, paragraph two might have been written like this:

Our investigation followed your lengthy correspondence with Mr. and Ms. Parsenon regarding their complaints of high gas consumption and cold floors. In your letter of February 16, 19xx, you authorized us to determine the extent of high fuel consumption, identify the cause, and suggest possible remedies.

This "personalization" of a report's language becomes even more noticeable when the person to whom a letter report is addressed is an individual citizen rather than a company. If Tod Phillips had addressed his report directly to Mr. and Ms. Parsenon, rather than to the construction company, its Summary would have been much more personal:

Dear Mr. and Ms. Parsonon:

We have investigated the reported high heating fuel consumption in your home, and have found it to be 4.37% higher than in homes of comparable age, size, and construction. Our checks revealed no faults in the heating system, or that your bungalow had been insulated inadequately during construction. Comparison with comparable homes, however, showed that the high gas consumption is probably caused by the higher-than-average temperature at which you maintain your home.

We believe you could reduce gas consumption to a normal or even lower level by maintaining a slightly lower temperature in your bungalow during the day, lowering the thermostat to about 65°F at night, and insulating the upper four feet of the basement walls. If you want to obtain even greater fuel savings, then we suggest you consider installing additional insulation in the ceilings and walls.

Our investigation was requested by . . . (etc.)

7
Suggestions and Proposals

Proposals vary from short memorandums to multivolume hardbound documents. Those discussed here are the shorter, less formal versions which an individual is more likely to write, ranging from a single page to probably five or six pages.

There are three types of proposal:

1. Informal suggestion
2. Semiformal proposal
3. Formal proposal

A **Suggestion** offers an idea and briefly discusses its advantages and disadvantages. (For example, a supervisor may suggest to a department manager that coffee break times be staggered, to avoid lineups at the mobile refreshment wagon.) Most suggestions are internal documents and are written as memorandums.

A **Semiformal Proposal** presents ideas for resolving a problem or improving a situation, evaluates them against certain criteria, and often recommends what action should be taken. (For example, a supervisor may propose to management that steps to taken to overcome production bottlenecks in the company's packing department; he or she might suggest introducing new packaging equipment, discuss various alternatives such as hiring additional staff or embarking on a training program, and then recommend the most suitable approach.) A semiformal proposal may be written as a memorandum or letter, or even in semiformal report format.

A **Formal Proposal** describes an organization's plans for carrying out a large project for a major client or the government. It is a substantial, often impressive document which describes in considerable detail what will be done, how and when it will be done, why the organization has the capability to do the work, and what it will cost. Such a proposal is often prepared in response to a Request for Proposal (RFP) and is almost always submitted as a bound book similar to a formal report. In an extreme case it may run to several volumes.

Writing compartments are illustrated and described for all three types of proposal, and examples (with accompanying comments) are provided for the informal suggestion and semiformal proposal.

INFORMAL SUGGESTION

The writing compartments for an informal suggestion are illustrated in Figure 7-1, which shows:

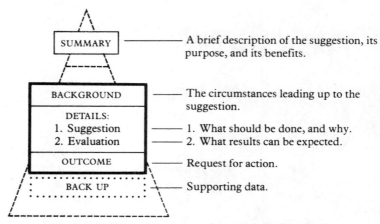

Fig. 7-1. Writing compartments for an informal suggestion.

- The **Summary** states very briefly what the proposer wants to do or wants done.
- The **Background** compartment describes the present situation.
- The **Details** section has two components:
  1. A *Suggestion* compartment, which outlines the suggested changes or improvements, and describes why they are needed.
  2. An *Evaluation* compartment, which identifies what effect the suggested changes or improvements will have, and categorizes them into advantages and disadvantages (sometimes called "Gains" and "Losses").
- The **Outcome** compartment identifies what action needs to be taken. It can either
  a. Request approval for the author to implement his or her suggestion; or
  b. Identify who should take the necessary action and possibly describe when it should take place and who should do the work.
- An optional **Backup** compartment contains supporting data such as cost estimates, records, plans, and sketches.

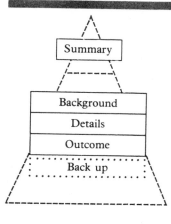

INFORMAL SUGGESTION:
Proposal for a Study

When an informal suggestion is as short as Anne Martin's, the writing compartments may not be as clearly evident as those used to shape a longer memo, letter, or report. Nevertheless they are still there, contributing to the overall organization of the suggestion.

1. Anne uses a "cause-and-effect" approach for her **Summary**, first very briefly identifying her reason and then stating in general terms what she would like to do.

2. This paragraph is the **Background** compartment; it amplifies the reason stated in the Summary.

3. The **Details** paragraph contains

   - A specific **Suggestion**, in sentence 1.
   - A short **Evaluation**, which spells out what is to be gained (sentence 2) and the cost (sentence 3). Anne keeps the evaluation short by referring to details in an attachment (**Backup**).

4. In the **Outcome** compartment Anne states specifically what she wants to do, and requests approval (**Action**).
   NOTE: To conserve space the letter from Brian Lundeen referred to as an attachment (the **Backup**) is not included here. His report of the computer analysis, however, is printed in Chapter 8.

THE WESTERN FARM IMPLEMENT COMPANY

Inter-office Memorandum

From: Anne Martin, Accountant   Date: October 13, 19xx
To: Frank Kelvin, President   Subject: Proposal for Computer Study

① Now that our business volume has reached a level where we probably should be employing a computer for accounting and inventory control, I propose we engage a consultant to identify our exact needs.

② Presentations made to me over the past 12 months by representatives of several computer companies have almost convinced me that we need a computer, but which system will best meet our current and future needs neither I nor anyone else in the company is qualified to evaluate.

③ I would like to obtain an objective analysis from Brian Lundeen of Antioch Business Consultants. As his attached letter describes, he will assess our existing and potential business volume and, if he believes we need a computer, will evaluate alternatives and recommend the best system. He has quoted a firm price of $850.00 to undertake such a study.

④ I suggest we authorize Antioch Business Systems to carry out the study, and request your approval to go ahead.

Anne

SEMIFORMAL PROPOSAL

The features that make a semiformal proposal more comprehensive than an informal suggestion are:

1. It usually deals with more complex situations, such as a problem or unsatisfactory condition.
2. It discusses the circumstances in more detail.
3. It establishes criteria (guidelines) for any proposed changes.
4. It frequently offers alternatives, rather than a single suggestion.
5. It analyzes the proposed alternatives in depth.
6. Its appearance is more formal.

The writing compartments for a semiformal proposal are shown in Figure 7-2. In more detail, they contain the following information:

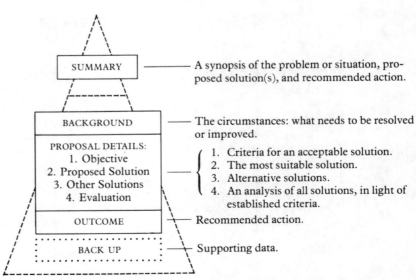

Fig. 7-2. Writing compartments for a semiformal proposal.

- The **Summary** briefly describes the main highlights of the proposal, mostly drawn from the **Background**, **Solutions**, and **Outcome** compartments. If headings are used in the proposal, this compartment is preceded by the word *Summary* or *Abstract*.
- The **Background** compartment introduces the problem, situation, or unsatisfactory condition, and outlines the circumstances leading up to it. It may be preceded by the heading *Introduction*.
- The **Objective** compartment defines what needs to be achieved to resolve the problem, and establishes criteria for an ideal, optimum, or best solution. This information may be included as part of the *Introduction* or preceded by a heading of its own (such as *Requirements* or *Criteria*).
- The two **Solutions** compartments describe various ways the problem can be resolved or the situation improved. Each alternative should include
  a. A description of the solution
  b. The result or improvement it would achieve
  c. How it would be implemented
  d. Its advantages and disadvantages
  e. Its cost

  Ideally, the alternative solutions will be arranged in descending order of suitability. These two compartments may be preceded by a single heading, such as *Methods for Increasing Productivity*, or by several descriptive headings.
- The **Evaluation** compartment analyzes and compares the alternative solutions, with particular reference to the criteria established in the **Objective** compartment. It may briefly discuss the effects of
  a. Adopting the proposed solution
  b. Adopting each of the alternative solutions
  c. Adopting none of the solutions (i.e., taking no action)
- The **Outcome** compartment recommends what action should be taken. It should be worded in strong, positive terms and, if headings are used, be preceded by the single word *Recommendation*.
- The **Backup** compartment, if used, contains detailed analyses, test results, drawings, etc., which support and amplify statements made in the previous compartments. It is usually preceded by the heading *Attachments* or *Appendices*.

The semiformal proposal on the following pages shows how these compartments can help organize information into a coherent, convincing, persuasive document.

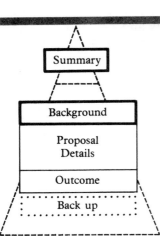

ANALYSIS OF SEMIFORMAL PROPOSAL: "Evaluation of Photocopiers" (Pp. 86–93)

1. The proposal's title is important: it should entice readers' interest. The key word here is *Improved*.

2. In a proposal the **Summary** should clearly identify the

   - Purpose (why a change is necessary)
   - Proposal (what the change involves)
   - Result (what is to be gained)

These should be highlights drawn from the **Background**, **Proposal Details**, and **Outcome** compartments.

82 Suggestions and Proposals

Often, it is better to write the Summary after the remainder of the proposal has been written (see the comments on Lorraine Dychuk's writing sequence in note 27).

The heading "Summary" is optional.

3. A Summary should encourage readers to read further, preferably by making the major gain(s) or advantage(s) readily apparent. A demonstrated cost saving always appeals to management.

4. The longer a proposal, the more need there is for headings. Those Lorraine uses help readers "see" her organization plan as they read. A list of the primary headings shows the logical sequence of her report:

```
Summary                          - (Main Message)
Existing Photocopier Facilities  - (Background or Situation)
Proposed Photocopier System    ⎫
Other Systems and Configurations ⎬ - (Proposal Details)
Evaluation of Alternatives     ⎭
Conclusions                    ⎫
                               ⎬ - (Outcome or Results)
Recommendations                ⎭
Attachments                      - (Backup Data)
```

5. The **Background** compartment starts here.

6. Rather than introduce a detailed cost analysis into the proposal narrative, it's better to place the details in an attachment (to become **Backup** data) and to include only the highlights or totals here. These highlights should sufficiently satisfy readers' curiosity so that they do not have to flip back to the attachment for more information.

7. This paragraph is an implied **Objective** compartment. (It's not always necessary to identify every writing compartment with a separate heading, although the longer a compartment the more likely it is to have one.) In the previous paragraph Lorraine has listed three deficiencies inherent in the existing photocopier system, and now implies that her proposal will both correct these deficiencies and achieve a cost saving.

8. Lorraine adds strength to her **Objective** compartment by identifying a particular aspect needing correction. In effect she is saying: "We must have a copier which can take oversize originals and reduce them to fit onto a standard-size page." Later she will demonstrate that only the proposed system can do this.

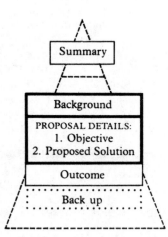

9. The **Proposal Details** compartment starts here with an analysis of alternatives. Normally the proposed solution (the author's "proposal") comes first and is discussed in detail. Alternative solutions follow, and are usually discussed in descending order of suitability. (Sometimes the sequence is reversed, but the reverse sequence is rarely as effective.)

Each alternative must be described factually and objectively. This is particularly important for the alternative the author proposes, because readers will automatically assume that the author favors it and is biased toward it. The analysis of each alternative should contain

83 Suggestions and Proposals

1. A description of the alternative (copier system, in this case), including what it is, what it can do, and why it is being considered
2. An analysis of its strong features, or advantages (what is to be gained if the alternative is adopted)
3. An analysis of its weak features, or disadvantages (what it will "cost", particularly if it will introduce negative side-effects)

10. This paragraph contains the initial description of the proposed alternative and refers the reader to two attachments where more comprehensive details may be found. Lorraine discusses costs in the following two paragraphs.

11. The advantages are discussed only briefly here, because Lorraine has already identified what features a new machine should have (at 7), mentioned that the proposed machine has them (at 9, 10), and discussed the cost advantages (in the two preceding paragraphs).

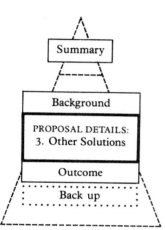

12. An analysis of the disadvantages starts here and continues for three paragraphs. It is much more detailed than the discussion of advantages because the author has to demonstrate that she has not only considered all the negative aspects, but has also found methods for overcoming or negating them. (Lorraine commented privately afterward that when she was writing her proposal she felt she must show readers she "left no stone unturned.")

13. Some authors prefer to write a short concluding statement at the end of each descriptive analysis, to sum up with a comment on the suitability of the particular alternative. Such a concluding statement is a reflection of the author's opinion, and hence has to be somewhat subjective. Other authors (and Lorraine is one of them) prefer to wait and present a combined descriptive analysis after they have described all the alternatives (see note 20).

14. & 15. Five other alternatives are considered, but only three have sufficient merit to warrant a detailed discussion. The two other alternatives are discussed briefly and then dismissed as being unnecessary or impractical, one at the beginning of this section (at 15), and one at the end (at 19).

16. The three viable but less effective alternatives are described in what is probably a decreasing order of suitability, although the margin between them is small. Each is given a heading so that the alternative can be readily identified. In each case Lorraine describes the alternative very briefly, discusses its cost, and mentions its main strengths and weaknesses. She maintains an objective, impersonal tone throughout.

17. Showing these tiny tables of cost calculations is a useful way to demonstrate the similarity of cost for all three systems, and for readers to compare each to the total cost of the proposed system.

18. The more remote or unlikely the alternative, the briefer the discussion and analysis may be.

19. See note 14.

84 Suggestions and Proposals

20. So far, Lorraine's descriptions of the possible systems have been direct, factual, and objective. Now she has to compare the systems and demonstrate that the one she proposes is better than any of the alternatives. At this point her style changes from descriptive to persuasive. Although she may be personally convinced that the PRO MK 3 is the "right" system, she still has to convince her readers that it is best so that they will agree with her conclusions when they read them, and accept her recommendation.

21. By setting criteria at the beginning of the **Evaluation** compartment, Lorraine provides a reference against which she can compare each alternative. If she had not first established the selection criteria she would have had to compare the alternative systems against each other, which would have been much more difficult to write briefly and convincingly.

The significance and validity of the selection criteria must be apparent to readers. It can be established here, or earlier in the proposal. (Lorraine established the reason for her criteria during her evaluation of the existing photocopying system.)

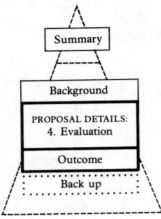

22. A consolidated comparison table like this lets readers "see" a comparison quickly and easily. It would take many words to say as much.

23. Lorraine makes sure that the expression "PRO MK 3" is noticed when each criterion is discussed. When more than one system meets the criterion, she always mentions PRO MK 3 first.

24. The **Conclusions** should evolve naturally from the proposal details which precede them. They should offer viable alternatives and must never introduce new information.

25. A person writing a proposal cannot afford to write a weak, wishy-washy recommendation. The **Recommendation** must be strong and convincing and clearly demonstrate that the author is personally sold on the course of action to be taken. Here is an occasion when the words, "I recommend . . ." must be used, not the impersonal, passive, "It is recommended that . . .," which seems to imply that the author is not quite sure of himself or herself.

26. There are three attachments to Lorraine's proposal:

1. A cost comparison
2. A comparison of copier features and capabilities
3. A brochure describing the proposed copier, which Lorraine obtained from the manufacturer's representative (Note: Attachment 3 has been omitted, because it is too bulky to include here.)

27. How Lorraine tackled this proposal, and particularly the sequence in which she wrote it, is worth commenting on because it demonstrates the usefulness of writing a comprehensive proposal or report *in reverse order* to that which readers encounter. This is what she did:

1. A major part of the project was a physical evaluation of copiers, during which she became convinced not only that an alternative system was necessary but also which system was the best. Consequently she knew she would

85 Suggestions and Proposals

be writing her proposal with its outcome clearly in mind, and hence could not avoid being somewhat subjective. Yet she also knew that to her readers the detailed analysis of copier systems must appear objective (unbiased) even though it would be leading up to a foregone conclusion.

2. Lorraine's second step was to develop the two attachments seen here. Preparing them at this early stage had a significant advantage in that it clarified her thinking. She had to identify the systems she would evaluate, the factors she would compare, and the categories she would place in each attachment (cost factors in one, and features and capabilities in the other). Doing all this drew the most important factors even more sharply into focus, and helped her organize the Proposal Details.

3. At this point she drew up a rough topic outline, from which the headings in the proposal eventually took shape. Lorraine's outline looked like this:

> SUMMARY: propose a system which will be better and cheaper.
> BACKGROUND: what we have now; plus what we haven't got and really should have.
> PROPOSED SYSTEM: PRO MK 3 (how it meets all needs).
> ALTERNATIVES:
> - 2 Fascopy
> - 2 PRO MK 1
> - 1 of each
> - Offset printer plus copier
>
> (the good and the bad of each)
>
> EVALUATION: establish criteria; compare systems to them.
> CONCLUSIONS: best and second-best systems.
> RECOMMENDATIONS: lease PRO MK 3; cancel existing leases.
> ATTACHMENTS: costs; features; brochure of PRO MK 3.

4. She then started writing, beginning with the Background and working steadily through the Proposal Details toward the Conclusions and Recommendations.
5. She had a draft typed, which she edited.
6. Finally, Lorraine wrote the Summary, selecting the key points from the draft proposal.

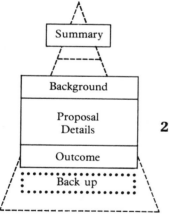

28. An attachment, and particularly a comparative analysis, provides evidence which supports statements made in the proposal. This evidence must appear unbiased, otherwise the proposal will lose credibility. Facts and figures provide the necessary objectivity, but subjective comparisons such as "fair," "good," and "very good" sound like unsupported opinions unless their source is identified. For the three existing copiers the author is the source and her opinions are based on experience. For the unknown machine she has referred to two independent sources, both having experience with the machine.

PROPOSAL FOR IMPROVED COPYING FACILITIES

by

Lorraine Dychuk
Systems Analyst

Summary

My evaluation of the company's existing photocopiers shows that by adopting an alternative system we could obtain better photocopying services at lower cost. To achieve this objective I propose that the company consolidate its copying services to a central location and exchange the four middle-range copiers we now use for one top-flight copier with much greater capabilities. The cost saving will be a minimum $122 per month, or 11.5% less than our current copying costs.

Existing Photocopier Facilities

The company has four photocopiers, each used in a different location. They are:

* Two leased Valiant "Fascopy" machines, one in the front office and one in the Production department.

* A leased Vancourt "PRO MK 1," which is shared by the Marketing and Information Systems departments.

* A company-owned Aetna "Little Gem," in the Purchasing and Supply department.

The three leased machines together cost $460 per month for their basic leases, plus an additional charge for each copy made. These "per copy" charges, when combined with the cost of materials such as paper and printing fluids, amount to a further $596 per month, for an average total monthly cost of $1056. A breakdown of these costs is included in attachment 1.

The four machines combined make about 34,000 copies a month, at speeds ranging from a slow 15 copies per minute to a moderately fast 40 copies per minute. They provide a reasonably good service at an average cost of 3.1 cents per copy. And yet none of the machines has the capability to:

* Print on both sides of the paper.

* Copy from oversize originals.

* Reduce the size of the copies.

These features, which are available on newer, more expensive machines, would be useful additions to our copying services. In particular, they would enable us to achieve mailing economies by printing reports on both sides of the paper, and to reduce oversize computer sheets to fit on a standard page.

Copying sheets of computer data has been a recurring problem. Either they must be retyped (a slow process, with the ever-present fear that errors may be introduced into the data), or copied in two passes, with the sheets subsequently taped together and folded into the report (also a slow process, with a cumbersome result).

Proposed Photocopier System

All the features we would like to have -- and more -- can be obtained by leasing an advanced-technology high-speed copier. The model I have investigated is the Vancourt Business Systems' PRO MK 3, which is a much-upgraded version of the PRO MK 1 currently used by the Marketing and Information Systems departments. It not only prints on both sides of the paper and copies from and reduces large-size originals, but also produces 60 copies per minute and has an optional 20-station automatic collator. Its features and capabilities are compared to those of our existing machines in attachment 2, and a brochure describing the PRO MK 3 is in attachment 3.

The PRO MK 3 leases for $360 a month (without collator). If it were to replace the four existing copiers it would assume a monthly workload of slightly less than 34,000 copies, and I estimate its additional features would probably encourage us to make about 2000 more copies per month than we make now. The per-copy charge and materials costs for this 36,000 copies would amount to $504, for a total monthly cost of $864, or 2.4 cents per copy. This is $192 per month, or 0.7 cents per copy, less than we are currently paying.

A 20-station automatic collator can be attached to the PRO MK 3 for $70 per month, resulting in a total monthly cost of $934, or 2.6 cents per copy. This is $122 per month, or 0.5 cents per copy, less than we are paying now. The cost breakdown in attachment 1 compares the current and proposed costs in greater detail.

(11) Having only one copier would mean consolidating all photocopying services in one location, which has both advantages and disadvantages. The advantages lie in more efficient copier management, better copier features, and lower operating costs. (Nearly all copier companies adjust their "per copy" charges according to the number of copies made each month: the more copies made, the lower the individual copy charge. For the PRO MK 3, the most significant price break occurs at 30,000 copies per month. Below 30,000, the charge is 1.7 cents per copy; above 30,000, it drops to 1.4 cents per copy.)

(12) The disadvantages of a single-copier operation are the lost convenience of near-the-job copying and dependence on only one copier. Departments which currently have their own copier are likely to resist the proposed consolidation, although most will quickly recognize the advantages of the expanded features an advanced-technology machine would bring with it. Dependence on a single copier, however, needs to be examined more carefully.

Three factors have to be considered: machine reliability, the speed with which the manufacturer responds to service calls, and the ready availability of an alternative service in an emergency. Our experience with the PRO MK 1 has shown us good machine reliability and that the local office of Vancourt Business Systems consistently responds to service calls within four hours. As a further check I contacted Mansask Insurance Company and Remick Airlines, both of which use the PRO MK 3, and found their experience to be the same with the new machine.

Standby services can be provided in two ways. For routine copying we can retain the Aetna Little Gem, which the company owns, and use it for making average-quality copies. For important jobs we can take the work to Cathcart Copy Corner on the Mezzanine of this building, which we already use for long-run high-quality printing. Ms. Cathcart has guaranteed to give our work priority attention in an emergency, providing we give her 30 minutes' notice. However, the need to use these standby services would be rare, since the speed of the PRO MK 3 would enable it to catch up quickly on backlogs
(13) following breakdown service or regular maintenance.

(14) Other Systems and Configurations

In the early stages of my evaluation I studied photocopiers made by other manufacturers to identify whether they had better features or price advantages. I found that, on average, capabilities and prices were similar, and
(15) that there seemed to be no significant advantage in switching to a machine made by another manufacturer. The quality, reliability, and service performance we have experienced with the three machines we now lease (see attachment 2) convinced me that we should stay with products we know.

To achieve both a cost saving and better management of the existing copying services will mean reducing the quantity of copiers we now have and consolidating those that remain into a "copying facility." The company-owned Little Gem should be withdrawn from service (other than being kept as a standby machine) because it is slow, costly, and produces copies on sensitized paper. And the remaining three copiers should be reduced to two, which would mean keeping only the two Fascopy machines, bringing in a second PRO MK 1, or retaining one of each. These alternatives are discussed below.

1. <u>Two Valiant Fascopy Copiers</u>

    This is the most economical alternative, resulting in a total monthly cost of $834, which represents a per-copy cost of slightly over 2.4 cents:

Basic lease (2 @ $145)	$290
Charge per copy (34,000 @ 1.6¢)	544
	$834

    The Fascopy provides high-quality copies on bond paper at a reasonably fast speed of 40 copies per minute. It does not, however, copy onto both sides of the paper, accept oversize images, or have a reduction capability. Neither can its copies be fed directly into an automatic collator. A summary of its features and capabilities is provided in attachment 2.

2. <u>Two Vancourt PRO MK 1 Copiers</u>

    Two PRO MK 1 copiers would cost $918 per month, or 2.7 cents per copy:

Basic lease (2 @ $170)	$340
Charge per copy (34,000 @ 1.7¢)	578
	$918

    The PRO MK 1 has an average speed of 30 copies per minute, copies onto only one side of bond paper, cannot accept large originals, and has no reduction capability. It can, however, be connected to a 10-station automatic collator, which can be leased for an additional $50 per month. With one collator, the monthly cost of two PRO MK 1s would increase to $968 or slightly more than 2.8 cents per copy.

3. <u>One Fascopy and One PRO MK 1</u>

    Retaining one of each copier type permits an automatic collator to be included in the configuration, but offers no further advantages. The cost would be $876 without a collator, or $926 with a collator, resulting in a per-copy cost of 2.6 or 2.7 cents:

Basic lease (1 @ $145; 1 @ $170)	$315
Charge per copy (17,000 @ 1.6¢)	272
(17,000 @ 1.7¢)	289
	$876

(19) I also briefly considered the feasibility of retaining only one copier (most likely the Fascopy) and running it in parallel with an offset press. I discussed using an offset press with Ms. Cindy Cathcart of the Cathcart Copy Corner, who agreed that it would be useful for doing professional quality printing at high speed (up to 9000 copies per hour), but pointed out that to achieve high quality press work demands expensive ancillary equipment and hiring or training someone to run the press and the equipment. After a tour of her printing facility I discarded the concept of acquiring our own offset press as being both impractical and uneconomical for the comparatively small volume of printing we would need to do.

(20) Evaluation of Alternatives

To compare the alternative systems I identified four features or capabilities a new system preferably should have, and two features which would be advantageous but are less essential. The more important features are:

1. A significant cost saving.
2. The ability to copy from oversize originals and reduce them to fit on a standard size page.
3. High system reliability coupled with fast service.
4. High operating speed.

(21) The two less essential features are:

5. Automatic feed to a collator.
6. The ability to print on both sides of the paper.

For a copier system without a collator, either the single PRO MK 3 or two Fascopy machines offer the lowest per-copy cost of 2.4 cents (see Table 1). With a collator, the lowest per-copy cost would be 2.6 cents, which is achieved only by the single PRO MK 3. These per-copy costs are considerably less than the current average per-copy cost of 3.1 cents.

TABLE 1: COST COMPARISONS FOR ALTERNATIVE SYSTEMS

(22)

System:	Existing	2 Fascopy	2 PRO MK 1		Fascopy + PRO MK 1		PRO MK 3#	
Monthly Cost:	$1056	$834	$918	$968*	$876	$926*	$864	$934*
Per-copy Cost:	3.1¢	2.4¢	2.7¢	2.8¢	2.6¢	2.7¢	2.4¢	2.6¢

* with collator

\# PRO MK 3 based on 36,000 copies per month; all others: 34,000 copies per month

Only the PRO MK 3 has the capability to copy from oversize originals. It can accept originals up to 17 x 11 in. (43 x 28 cm), and can reduce any original by as much as 50%.

All three machines have satisfactory reliability records and good service reputations.

For the number of copies we make or are likely to make each month, a copier speed of 2400 copies per hour (40 copies per minute) would be acceptable. Both the PRO MK 3 and the Fascopy operate at this speed or better.

Only the Vancourt PRO machines accept automatic collators. The PRO MK 3 has an optional 20-station collator; the PRO MK 1 has an optional 10-station collator, which would be of limited use for the marketing manuals we assemble.

Only the Vancourt PRO MK 3 can copy onto both sides of the paper.

Conclusions

Consolidating our photocopier system can create considerable operating economies and provide an overall improvement in service. Of the four systems I evaluated, a single Vancourt PRO MK 3 can offer all the essential features and capabilities we desire. A two-copier Valiant Fascopy system would offer the same economies but would not be able to accept or reduce large-size originals such as computer printouts, or print on both sides of the paper.

Recommendations

I recommend that the company:

1. Leases a PRO MK 3 photocopier from Vancourt Business Systems Inc.

2. Terminates the leases on the three currently leased photocopiers.

3. Withdraws the company-owned Aetna Little Gem from service, and retains it as an emergency standby copier.

ATTACHMENT 1

COST COMPARISON: FOUR EXISTING COPIERS VS PROPOSED SINGLE COPIER

LOCATION	MODEL	MONTHLY RENTAL	COPIES PER MONTH	X	COST PER COPY*	=	COPY COST PER MONTH	TOTAL COST/MONTH	TOTAL COST/COPY
Existing Copiers									
Front Office	Valiant Fascopy	$145	13,500		1.6¢		$216	$361	2.67¢
Marketing & Information Systems	Vancourt PRO MK 1	$170	8,600		1.7¢		$146	$316	3.67¢
Production	Valiant Fascopy	$145	9,000		1.6¢		$144	$289	3.20¢
Purchasing & Supply	Aetna Little Gem	--	2,800		3.2¢		$90	$90	3.20¢
		$460	33,900				$596	$1056	3.10¢ (ave)
Proposed Copier									
(Location to be selected)	Vancourt PRO MK 3 + Collator	$360 $70	36,000#		1.4¢		$504	$864 $70	2.40¢ 2.60¢
		$430	36,000				$504	$934	

* includes materials and cost-per-copy charge
\# estimated; based on better facilities creating increased usage

ATTACHMENT 2

COMPARISON OF COPIER FEATURES

Features	Valiant FASCOPY	Vancourt PRO MK 1	Aetna LITTLE GEM	Vancourt PRO MK 3
Speed (copies/min)	40	30	15	60
Maximum size of original accepted	14 x 8½ in. (36 x 21.5 cm)	14 x 8½ in. (36 x 21.5 cm)	13 x 8 in. (33 x 20 cm)	17 x 11 in. (43 x 28 cm)
Copy quality	Very good	Very good	Good	Excellent*
Uses bond paper	Yes	Yes	No	Yes
Image reduction	No	No	No	Yes; down to 50%
Copies on both Sides of paper	No	No	No	Yes
Copies all colors	Yes	Yes	No; not light blue	Yes
Accepts a collator	No	Yes (10-station)	No	Yes (20-station)
Reliability record	Fair	Good	Very good	Good*
Fast maintenance service (4 hours)	Yes	Yes	No (2 days)	Yes*
Manufacturer	Valiant Office Products	Vancourt Business Systems Inc	Aetna Copiers Inc	Vancourt Business Systems Inc

* Opinions of two users: Mansask Insurance Company and Remick Airlines

THE FORMAL PROPOSAL

Formal proposals are normally lengthy documents which sometimes run into several volumes. Hence, their size prohibits a sample from being included here. Instead, a typical outline and the purpose of each compartment are described below.

Most formal proposals are written in response to a Request for Proposal (RFP) issued by the government or a large commercial organization. Normally time is short between the date an RFP is issued and the date the proposal must be presented to the originating agency. Each company submitting a proposal forms a team of key individuals, who work many late hours to ensure their proposal is written, illustrated, printed and delivered before the closing date.

Many agencies issuing RFPs stipulate the major topics the proposing company must address and the sequence in which information is to be presented. Unfortunately, although there is some similarity between the formats stipulated by different agencies, there are sufficient variances to make it impossible to present a "standard" outline here. The outline illustrated below, therefore, is a simplified composite of several outlines, and is applicable to either a solicited or unsolicited (i.e., company-initiated) proposal.

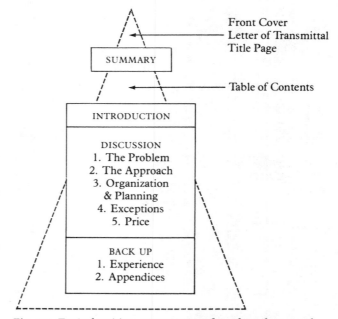

Fig. 7-3. Typical writing compartments for a formal proposal.

The major components of a proposal are illustrated in Figure 7-3. When the compartments are converted to headings, the outline looks like this:

Cover
LETTER OF TRANSMITTAL
Title Page
SUMMARY
Table of Contents
INTRODUCTION
DESCRIPTION OF PROBLEM or SITUATION
APPROACH TO RESOLVING PROBLEM or IMPROVING SITUATION

```
ORGANIZATION AND PLANNING
EXCEPTIONS
PRICE PROPOSAL
EXPERIENCE
    COMPANY
    EMPLOYEES
APPENDICES
```

The major components are shown in capital letters and are discussed briefly below.

Letter of Transmittal. When attached to a formal proposal, a Letter of Transmittal assumes much greater importance than the standard cover letter pinned to the front of a semiformal or formal report. Normally signed by an executive of the proposing company, it comments on the most significant aspects of the proposal and sometimes the cost. As such it has a similar role to the Executive Summary which sometimes precedes a formal report.

Summary. The Summary mentions the purpose of the proposal, touches briefly on its highlights, and states the total cost. If a letter of transmittal is bound inside the report, the Summary is sometimes omitted.

Introduction. As in a report, the Introduction describes the background, purpose, and scope of the proposal. If the proposal is prepared in response to an RFP, reference is made to the RFP and the specific terms of reference or requirements imposed by the originating authority.

Description of Problem or Situation. This section describes the problem that needs to be resolved or the situation that needs to be improved. It usually includes: (1) a statement of the problem/situation, as defined by the RFP; (2) an elaboration of the problem/situation and its implications (to demonstrate the proposer's full comprehension of the circumstances); and (3) the proposer's understanding of any constraints or special requirements.

Approach to Resolving Problem or Improving Situation. The proposer describes the company's approach to the problem/situation, and states specifically what will be done and why it will be done, and then in broad terms outlines how it will be done. As this is the proposer's solution to the problem or method for improving the situation, this section must be written in strong, definite, convincing terms which will give the reader confidence that the proposing company knows how to tackle the job.

Organization and Planning. Here, the "how" of the Approach section is expanded to show specifically what steps the proposer will take. Under "Organization" the proposer describes how a project group will be established, its composition, its relationship to other components of the company, and how it will interface with the client's organization. Under "Planning" the proposer outlines a complete project plan and, for each stage or aspect, exactly what steps will be taken and what will be achieved or accomplished.

Exceptions. Sometimes a company may conceive an unusual approach which not only solves the problem but also offers significant advantages, yet deviates from one or more of the client's specified requirements. These "exceptions" are listed and the reason why each need not be met is clearly explained.

Price Proposal. The proposer's price for the project is stated as an overall price and broken down into schedules for each phase of the project. The extent and method of pricing is often specified by the RFP.

This section of the proposal is the one most likely to be found in varying positions. The RFP may stipulate that it appear at the front, here, as the last section, or even as a separate document.

Experience. The proposing company describes its overall experience and history, and particular experience in resolving problems or handling situations similar to that described in the RFP. Key persons who would be assigned to the project are named, and their experience is described in curricula vitae.

Appendices. The appendices contain supporting documents, specifications, large drawings and flow charts, schedules, equipment lists, etc., all of which are referenced in the proposal.

Proposal Appearance. Major proposals are multipage documents assembled into book form and normally bound by a multiring plastic binding. Minor proposals have fewer pages but are still bound or stapled into book form. Some very short proposals, particularly those submitted from one company to another, may be simply stapled together like a semiformal report, or even in some cases submitted as a letter.

IV
FORMAL REPORTS

8
The Formal Report

Formal reports have a much more commanding presence than informal or even semiformal reports. Normally bound with a simple but dignified jacket, they immediately create the appearance of an important document. Internally, their information is compartmented and carefully spaced to convey a confident impression from start to finish.

The term *formal report* refers to the type of document rather than its title, and never appears on the title page. A formal report is more likely to be referred to more specifically as a

Feasibility Study
Investigation or Evaluation Report
Product Analysis, or
Project Report

Sometimes one of these names may precede the report's main title, but more often the name is omitted and the title stands alone.

TRADITIONAL ARRANGEMENT OF REPORT PARTS

There are six main compartments in a formal report, the initial letters of which form the acronym SIDCRA (see Figure 8-1):

Summary
Introduction
Discussion
Conclusions
Recommendations
Appendix

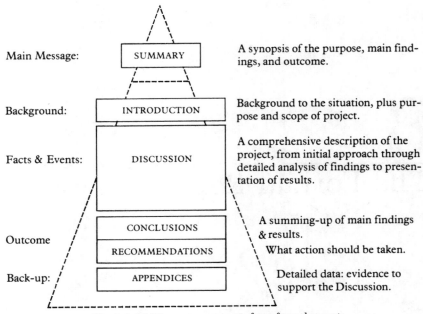

Fig. 8-1. Writing compartments for a formal report.

The first four compartments are identical to the four basic compartments identified in Chapter 2. If the writer of a formal report also proposes that action be taken, then the report has a Recommendations compartment. And if it contains supporting data, it has an Appendix.

The six main report parts identified in Figure 8-1 are the primary information-bearing compartments of a formal report. Additional parts support these main compartments and help give the report its formal shape. They are listed below in their appropriate positions in relation to the main compartments:

Cover Letter
Cover Page
Title Page
SUMMARY
Table of Contents Page
INTRODUCTION
DISCUSSION
CONCLUSIONS
RECOMMENDATIONS
References or Bibliography
APPENDIX
Back Cover

Guidelines for writing these parts appear on the following pages, facing the relevant compartments of Brian Lundeen's computer systems evaluation.

ALTERNATIVE ARRANGEMENT OF REPORT PARTS

To meet the needs of executive readers, sometimes the Conclusions and Recommendations are brought forward so that they follow immediately after the Introduction. This rearrangement helps readers gain a more complete picture of the report's outcome without having to delve into the deeper details of the Discussion. Coincidentally, it fully completes the pyramid structure, as shown in Figure 8-2.

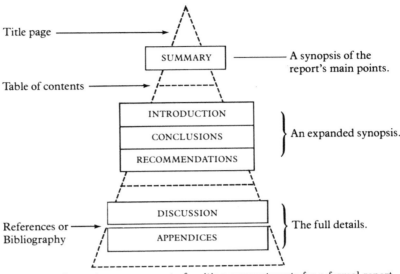

Fig. 8-2. Alternative arrangement of writing compartments for a formal report.

Moving the Conclusions and Recommendations forward does not materially affect report content, but does demand careful attention by report writers. They must ensure that

1. The transition from Introduction to Conclusions is achieved smoothly, naturally, and coherently
2. The Conclusions and Recommendations can be clearly understood (report writers have to remember that their readers will not yet have read the Discussion)
3. The Discussion does not start and end too abruptly (it is no longer preceded immediately by the Introduction or followed by the Conclusions). The Discussion may need a new introductory paragraph, and an additional paragraph or two to conclude the report.

ANALYSIS OF A FORMAL REPORT (Pp. 113–30)

COVER LETTER. A cover letter is really a transmittal document used to convey a report from one organization to another (or one person to another). The letter is paper-clipped to the *outside* of the report's front cover (in effect, the report is attached to the letter), and is not part of the report.

A cover letter can be a simple statement:

I enclose our report No. E-28, "Evaluation of Computer Systems for The Western Farm Implement Company," which has been prepared in response to your request of November 2, 19xx.

Alternatively, it can offer a miniature summary, as Brian Lundeen's cover letter does.

Occasionally, a "Letter of Transmittal" is used instead of the cover letter. A letter of transmittal is considerably more detailed, in that it may discuss aspects which are of particular importance to management, such as financial considerations, or it may draw attention to unusual factors affecting the project and the ensuing report. A letter of transmittal is always bound *inside* the report cover, so that it becomes an integral part of the report rather than a separate document.

TITLE PAGE.   The title page contains four main elements:

1. The full title of the report, which must be informative without being too long. Brian's title is informative; a very brief title such as "Computer Evaluation" would not be informative.
2. The name of the organization and sometimes the person for whom the report has been prepared.
3. The name of the originating organization, and sometimes the name of the person who has written the report.
4. The date the report is issued. If a report number also is included on the title page, it and the date are offset left and right as shown. If there is no report number, the date is moved to the center.

A title page must have eye appeal yet be simple and dignified. Every line must be centered individually on either side of a vertical centerline, which is offset about one-third of an inch to the right of page center to allow for the unusable ½ to ¾ inch on the left-hand edge of each page (see Figure 8-3). This edge usually is covered by the binding or punched for multiple-ring binders. If a report is printed on both sides of the paper, the binding edge on the reverse side of each sheet is on the right-hand edge of the page and the centerline is offset about one-third inch to the left.

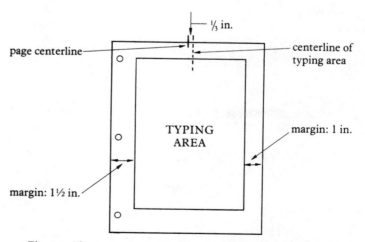

Fig. 8-3. The typing center line is offset from the page center line.

SUMMARY. In a formal report the Summary follows immediately after the title page but before the table of contents. It always has a page to itself, and is centered on the page. Sometimes it is preceded by the heading ABSTRACT instead of SUMMARY.

The Summary is the most important page in the report. Because it is the first body of information that readers encounter, it has to encourage them to read further; if it does not, it has failed to achieve its purpose.

Some guidelines for writing summaries are:

1. Write the Summary after the rest of the report has been written, but place it at the front of the report.
2. Draw information for the Summary from
    - The Introduction (particularly the purpose of the project)
    - The Discussion (pick out the most important highlights)
    - The Conclusions and Recommendations (the outcome or result of the project)
3. Keep the Summary as short as possible and make it interesting and informative. For example, rather than write, "Conclusions are drawn and a recommendation is made," state specifically what the main conclusions are and what you are recommending or suggesting should be done.
4. Keep the intended reader(s) clearly in mind, to ensure that you tell him, her, or them what they most want to know or need to hear.
5. Use plain, nontechnical words, and avoid topic jargon, so that the Summary can be understood by almost any person who picks up the report. (Notice how Brian Lundeen has avoided using computer terminology in his Summary.)

You will probably revise the Summary more than any other part of a report.

TABLE OF CONTENTS. A Table of Contents (T of C) may seem to be inserted at the front of a report mainly to help readers find specific information. But it also has a secondary, much more subtle purpose: that is, to let readers see how the author has organized the information and what topics are covered.

Readers who have read no more than Brian Lundeen's Summary can quickly establish his approach from the T of C. They will notice the logic of his organization: how he starts by comparing the existing (manual) system with computer systems generally, describes computers and their parts, and establishes factors to be considered ("Evaluation Considerations"); then he discusses the two main types of computer, and divides the discussion of each into two secondary topics; and, finally, he compares the costs and primary advantages and disadvantages of each system. Knowing this arrangement in advance helps readers adapt more readily to the information presented in the Discussion which follows.

Some factors you should take into account when writing a T of C:

1. Every major topic heading in the report must also appear in the T of C.
2. The topic headings in the T of C must be worded exactly as they appear in the report.
3. Minor subordinate headings may be omitted from the T of C if their inclusion would make the T of C too lengthy or detract from the clarity of the overall organization plan.

4. All appendices must be listed, with the complete title drawn from the first page of each appendix.
5. If drawings or illustrations are grouped separately in the report, they should be listed in the T of C. If there are many illustrations, it is acceptable to insert the single entry "Illustrations" and page number in the T of C, and to place a separate list of illustrations as the first page of the illustration section.

1. INTRODUCTION. The Introduction prepares readers for the details that follow in the Discussion. It introduces them to the circumstances leading up to the project, and the reason it was undertaken and the report was written.

   There are three main components:

   1. **The Background**, which describes events leading up to the existing situation, what projects (if any) have been done previously, and why the project or study is necessary.
   2. **The Purpose**, which defines what the project or study is to achieve, who authorized it, and the specific terms of reference.
   3. **The Scope**, which outlines any limitations imposed on the project, either by the person(s) authorizing it or by the person(s) undertaking it, such as cost, time in which the project is to be completed, depth of study, and factors which must be included or may be omitted.

   In very long reports these components may be treated as separate topics and even be preceded by individual headings. More often, however, they are interwoven and treated in whatever order the report writer finds most convenient. For example, in Brian Lundeen's report the background is in paragraph 2, the purpose is implied in paragraph 1 and stated specifically in paragraph 2, and the scope is in paragraph 3.

2. DISCUSSION. The Discussion starts here. Brian opens this section of the Discussion with a summary statement which immediately answers WFIC's question: "Do we need a computer?" The remainder of the section provides evidence to support his statement. This application of the pyramid method of writing to individual sections is an excellent organizational technique.

   Note that:

   1. The word "Discussion" seldom appears as part of a heading, and is never used as a single-word heading.
   2. The Discussion may follow immediately after the Introduction (i.e., on the same page), or start on a fresh page. If it starts on a fresh page, it's customary also to start the Conclusions on a fresh page.

   Because readers like to know where they are going, a report writer must make his or her approach readily apparent to readers before they encounter too many details. The approach normally is stated at the beginning of the Discussion, but may be included with the Introduction. Brian has chosen to combine his approach with the "scope" section (it's in the third paragraph of his Introduction) to enhance reading continuity and avoid repeating information.

   How you arrange the information within the Discussion is extremely important. The overall logic of the case you present must be clear to readers, otherwise they may follow your line of reasoning with some doubt or hesitation. This, in turn, may affect their acceptance or rejection of your conclusions and recommendations.

Three factors can have a particularly negative effect on readers:

1. Writing which is beyond their comprehension; that is, uses technical terms and jargon they may not understand
2. Writing which fails to answer their questions or satisfy their curiosity; that is, does not anticipate their reactions to the facts, events, and concepts you present
3. Writing which either overestimates or underestimates the readers' knowledge; that is, assumes they know more (or less) about the topic than they really do

These common pitfalls can be avoided if you clearly identify your readers. You have to establish first whether your reports will be read primarily by management, by specialists knowledgeable in your field, or by lay persons with very limited knowledge of your specialty. Then you must decide what they are most interested in, and what they need to hear from you (in case there is a conflict between what they would like to hear and what you need to say). And, finally, you have to plan your report so that the order in which you present information will answer not only their immediate questions but also any questions generated while they read.

For a start, go back to the terms of reference you were given by the person or organization authorizing your project. Pick out the points of most interest to your reader(s), jot them down, and then rearrange them in a logical sequence which will satisfy the readers' curiosity in descending order of importance.

As a business consultant, Brian Lundeen is fairly knowledgeable about computers. But he has to recognize that for this report his readers will not be knowledgeable. Who are they? Principally, Anne Martin and Frank Kelvin, accountant and president of WFIC. Other members of the company may also read the report, but these two will decide whether a computer is to be introduced into the firm.

What do they most want to know? They have two primary questions:

1. Do we need a computer?
2. If so, which one?

Brian answers question 1 in the first paragraph of the Discussion. Question 2 is answered by the remainder of the report.

3. Because Brian must use some computer terminology in his report, he takes time to describe computers in essentially simple terms, before readers encounter them. The first paragraph of this section is a summary statement (the top of the pyramid), and is intended to calm readers' fears that they are entering a highly technical domain. Note that although Brian's language is simple, it does not "talk down" to readers: he still uses terminology any competent business person should understand.

4. By introducing the words "hardware" and "software" in the second sentence of the section summary, Brian is subtly saying to his readers: "This is the organizational plan I will adopt for successive paragraphs of this section." His headings substantiate that statement.

5. Brian limits his definitions to those terms he will use when describing computer systems. There are many other terms he could have defined, but he has not done so because too many technical terms might have confused his readers. For example, it would be natural for Brian to use the computer term "bytes" as a measurement of memory capacity; he has chosen to use "bits" instead, because the word would be more readily recognized by his readers and does not need definition.

6. Illustrations such as drawings, sketches and photographs provide a useful way to help readers visualize complex topics. But they should be chosen and inserted with care, according to the following guidelines:

  1. They should always serve a useful purpose.
  2. They must supplement, not duplicate, the written words.
  3. They must be simple, clear, and readily understood.
  4. They must be referred to in the narrative of the report.
  5. They should be accompanied by a brief caption or title, and sometimes a few explanatory remarks.
  6. Ideally, they should be smaller than full-page, so that some text can appear either above or below them (full-page illustrations tend to interrupt reading continuity).

More detailed information, particularly for preparing charts, graphs, and tables, is provided in Chapter 12.

7. As this is one of the criteria Brian will use to select the most suitable computer system, he must make sure readers are aware of it before he starts making comparisons. Note that this is an author-imposed criterion, as opposed to the client's criteria listed in the following paragraphs.

8. An author should establish any guidelines that influence the project as early as possible in the report. Readers need to know about them so they will better understand the author's approach and the course the project took. Such knowledge also helps precondition readers to more readily accept the project outcome.

9. This is an unfortunate arrangement: when a colon introduces a list, it should never be the last item at the foot of a page. Brian should have had this page retyped and the last two lines moved to the top of the next page, where they could properly introduce the list that follows.

10. When a list can be read easily or is short, as this one is, it may be included in the report narrative. But if the list is long, or contains part numbers or coded items which make uninteresting reading, it should be placed in the appendix and a reference made to it in the report narrative. This reference should also comment on any data of particular significance. Here's an example:

> Materials required for the project, and their sources of procurement, prices, and delivery dates, are listed in Appendix B. Five items are coded LLT to indicate they have "long lead times" and must be ordered within 20 days of project approval.

11. This is another section summary statement, intended to orient readers before they start reading the details further on in the section. From it they will learn what the section is about and how the information is organized.

12. (See report page 7.) This is the first of several computer systems Brian describes and which he compares later on in the report. When he wrote these sections he had already decided which system would be best for his client, yet for each he has taken care to present the information objectively (without bias).

To maintain a parallel structure throughout, for each system description he has used a three-stage plan:

>INTRODUCTION
>FACTS
>EVALUATION

Each system description thus becomes a separate minireport within the overall structure of the whole report. This two-level "report-within-a-report" arrangement is used by many experienced report writers.

The compartments in this subsection of Brian's report are easy to identify:

13. The first two sentences are the **Introduction**.

14. The remainder of the first paragraph presents **Facts**.

15. The following two paragraphs **Evaluate** the system (they comment on its advantages and disadvantages).

16. Extensive details, such as manufacturers' catalogues, tabulated data, calculations, specifications, and large drawings, are placed in an appendix and stored at the back of the report where they will not interfere with reading continuity. They constitute evidence which supports and amplifies what is said in the body of the report.

When selecting and referring to appendices, a report writer must ensure that

1. They are necessary and relevant.
2. Every appendix is referred to in the report narrative.
3. Readers do not have to refer to the appendices in order to understand the report.

To prevent readers from having to flip pages and refer to appendices as they read, a report writer may have to include a synopsis of an appendix's highlights in the report narrative or draw a conclusion from the appended data. For example,

> A survey conducted in the Camrose shopping center (see appendix K) showed that during weekdays 74% of shoppers traveled by private automobile, and 23% by bus. On Saturdays the number of automobile travelers increased to 83%, and bus travelers decreased to 15%. The remaining 2–3% traveled by bicycle or on foot.

(Appendix K contained seven pages of numerals; if placed in the report they would have both physically and psychologically interrupted the narrative.)

17. The second computer system analysis starts here and its format is the same as that for the previous analysis:

18.   Introduction   (starts at 17)
Facts   (starts at 18)
Evaluation   (starts at 19)

20. Note that no comparisons are made between systems at this stage of the report: Brian's discussion is completely objective in the Introduction (17) and presentation of Facts (18). But in the Evaluation paragraphs (15 and 19) he cannot avoid letting subjectivity show to some extent (principally because he has been doing the evaluating, and so many comments he makes reflect his opinions). Consequently he has taken care only to make comparisons against the criteria he established earlier in his report, and not between one system and another.

21. This section of Brian's report demonstrates not only the "report-within-a-report" method described earlier, but also a third-level "report-within-a-report-within-a-report" technique. The section headings and compartments show how this has been done:

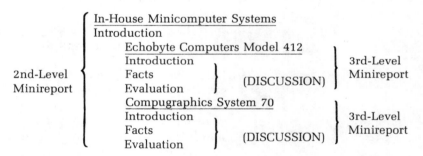

Note, however, that Brian has omitted the Conclusions compartment from the basic report structure. He explained that for most reports he would normally sum up each description with a short concluding statement, which to some extent would show how he felt about the suitability of the system. In this instance he did not want to prejudice his readers' opinions until they had had a chance to read and think about all the systems. Consequently he chose to write only the Introduction and Discussion in each minireport, and to save the Conclusions compartment until he had described and evaluated all the systems.

Although Brian's approach is seemingly nonstandard, it still successfully achieves its purpose. It also demonstrates that report writers should not feel they are entirely limited to a "standard" structure. They should be willing to experiment and adopt an alternative approach if to do so will result in a more effective report.

22. A new minireport starts here with an introductory statement which describes the timesharing system in broad terms. By defining it as having two main components ("external" and "in-house"), Brian prepares his readers to expect the ensuing narrative to be similarly divided.

23. The description of the "external" facilities in this section, and the "in-house" components in the next section (it starts at the next heading), together form the **Facts** and **Evaluation** compartments of this minireport.

24. Each paragraph contains both **Facts** and **Evaluation**. Generally, Brian introduces a fact at the start of each paragraph, and then evaluates it in the remainder of the paragraph.

25. Readers want not only to be given facts, but also to know how they are derived. If it is not practicable for a report writer to provide an in-depth discussion of certain facts (possibly because they would be irrelevant to the report's main thrust or divert readers' attention), then reference must be made to the source from which the information was obtained. Such references are grouped and placed at the end of a report, where they are called "References," or simply "Endnotes."

The raised "3" in the middle of line 5 refers readers to the third item listed in the References on report page 12.

(NOTE: Footnotes, which once were popular, are not used in modern reports, partly because they distract a reader's eye but mostly because making room for them at the foot of a page creates typing problems. See Chapter 11.)

26. The information describing the three types of computer terminals is mostly **Facts**, with an occasional comment or **Evaluation**. As before, Brian makes no attempt to compare one set of terminals with another, except for a very mild cost comparison.

27. Although Brian's typist tried to arrange the typing so that it would start and finish at the same level on every page, both here and on report pages 3 and 11 this objective could not be achieved. On page 3 the diagram which follows the last line of type is too large to be placed at the foot of the page and consequently had to be placed at the top of page 4; on page 9 only the heading and the first line of type (now at the top of report page 10) would have fitted on the page; and on page 11, if the recommendations had started immediately beneath the conclusions there would have been room for only the primary recommendation, which would have separated it from the secondary recommendations.

A heading should never stand alone at the foot of a page, and whenever possible a single line of type should not be placed at the foot or top of a page.

28. Having presented the facts about the various computer systems, Brian now starts comparing the available alternatives to help his readers understand and accept the conclusion he will shortly draw and the recommendations he will make. A comparative analysis like this must always be part of the report's Discussion, *never* part of the Conclusions.

Brian breaks the comparative analysis into two segments, each preceded by a main heading, and treats costs separately because they are probably of more importance to his readers than any of the other factors he evaluates. In a comparative analysis in which costs are less nebulous and only one of several factors of equal weight, there would be no need to segregate them.

29. Brian places the details in an appendix and presents only the total costs in the analysis. To include a detailed cost analysis here would interrupt continuity because readers would be tempted to stop and study it. If they were to start comparing individual details at this point their attention could easily be deflected from the report's main thrust.

30. A comparative analysis should always present the alternatives in either ascending or descending order of suitability; that is:

- In *ascending* order the least likely choice appears first and the most likely choice (the alternative the report writer will recommend) appears last.
- In *descending* order the best choice appears first and the least likely choice is placed last.

This "worst-to-best" or "best-to-worst" arrangement is a very subtle way to prepare readers for the final outcome of a comparative analysis. Brian uses the worst-to-best arrangement but does not tell his readers why he has presented the four systems in that particular order; he allows the arrangement to subtly prepare them to expect the conclusions he draws.

31. Report writers who have done a thorough project or study will have a clearly defined project outcome in mind. If they are to convince their readers that the outcome they describe is valid and the action they suggest is sound, eventually they must cease being purely factual and start persuading their readers to agree with their results. At this stage they can let their subjectivity show and their opinions become apparent.

32. The ascending order of presentation used in the previous section continues here: Brian first discusses the less preferable choice (a purchased in-house system) and then discusses the preferred choice (a leased system).

33. CONCLUSIONS. The Conclusions and Recommendations are sometimes referred to jointly as a "terminal summary," meaning that they provide a summing-up of the outcome of the Discussion. It's unwise, however, to treat them both under a joint heading, beause doing so can invite a report writer to inadvertently write a weak recommendation.

The most important thing to remember about Conclusions and Recommendations is that they must never offer surprises. That is, *they must present no new information*: everything they contain must have been discussed in previous sections of the report (i.e., in the Discussion). Introducing a new idea or thought into the Conclusions or Recommendations is one of the common faults found in business reports.

The Conclusions and Recommendations can either follow directly after the end of the Discussion, as shown here, or start on a fresh page. If the Introduction has a page (or pages) to itself, then the Conclusions should start on a new page.

The Conclusions should

1. Be as brief as possible, with their main points drawn from the concluding paragraph or statement of each section (or minireport) in the Discussion.
2. Be presented in descending order of importance: i.e., primary conclusion first, followed by subsidiary conclusions
3. Satisfy the requirements established in the Introduction
4. Never advocate action
5. Be presented in "point form" (in numbered subparagraphs) if there are many subsidiary conclusions

34. RECOMMENDATIONS.   The Recommendations should

1. Be strong, and advocate action
2. Use the active voice (They should say, "I recommend . . ." if you are personally making the recommendations, or, "We recommend . . ." if you are making recommendations for a group of people, a department, or your organization or company. They should never use the weak passive voice, as in, "It is recommended . . . ."
3. Satisfy the requirements established in the Introduction
4. Follow naturally from the Conclusions
5. Offer recommendations either in descending order of importance, or in chronological sequence if one recommendation naturally follows another
6. Be in point form when several recommendations are being made

35. REFERENCES/BIBLIOGRAPHY.   A list of references or a bibliography lists all the documents the report writer used while conducting the project. Each reference describes the source of a particular piece of information, and in sufficient detail so that readers can identify and obtain the document if they want to refer to it.

These are the main differences between a list of references (which Brian uses) and a bibliography:

1. References are numbered and appear in the sequence in which each piece of information is referred to in the report.
2. Bibliography entries are not numbered and appear in alphabetical sequence of authors' names.

Generally a list of references is more common in business and technical reports, and bibliographies are seen more often in professional journals and academic theses.

Some suggestions for writing a list of references or a bibliography are contained in Chapter 11.

36. APPENDIX.   The Appendix contains complex analyses, statistics, manufacturers' data, large drawings and illustrations, photographs, detailed test results, cost comparisons, and specifications—indeed, any information which if included in the Discussion would interrupt reading continuity. Often the appendix will contain detailed evidence to support what is said more briefly in the Discussion. And sometimes the appendix section contains more pages than all the remaining sections of the report put together.

112 *The Formal Report*

Certain guidelines apply to appendices (or "appendixes"—either plural is correct):

1. Appendices always appear in the order in which they are first referred to in the report (and, of course, every appendix must be referred to).
2. As appendices are considered individual documents, each may be page-numbered separately, starting at "1."
3. Each appendix is assigned an identifying letter: "Appendix A," "Appendix B," etc.

NOTE: To save space here, appendices A, B, and C have been omitted from the report because they contain numerous pages of manufacturers' catalogues and specification sheets. Appendix D has been retained to show how a two-page appendix containing author-originated data should be prepared.

37. Because each appendix is a separate document, sometimes an appendix will contain its own references. The references may be listed at the end of the appendix or treated as a footnote, as shown here.

38. This cost analysis supports the section of the Discussion titled "Comparative Costs of the Systems" and the reference to total costs mentioned in the Conclusions. It also consolidates in a single table all the comments on component costs mentioned throughout the report. If readers were to encounter this table in the Discussion they would want to stop and examine it, which would direct their attention *away* from the report proper. Consequently Brian places it in an appendix, where it can be examined later.

39. Large, wide, horizontal illustrations and tables which occupy a full page are always positioned so that they are read *from the right*. (See Chapter 12.)

40. It is *correct* for these titles to appear inverted. (See Figure 12-16, Chapter 12.)

ANTIOCH BUSINESS CONSULTANTS

3870 Spruce Drive, Foothills, Colorado 80409

December 18, 19xx

Anne D. Martin, Accountant
The Western Farm Implement Company
414 Mountain Highway
Foothills, CO 80409

Dear Ms. Martin:

I have examined The Western Farm Implement Company's records as requested in your letter of November 2, 19xx, and believe you should convert to a computerized system of accounting and inventory control. The enclosed report examines several systems, and suggests that initially you should lease time from a computer service company rather than purchase a complete system of your own.

Please call me if you would like to discuss the report or introduction of a computer into your operations.

Sincerely

Brian E. Lundeen

Brian E. Lundeen
Computer Consultant

BEL:cs
enc

ANTIOCH BUSINESS CONSULTANTS

EVALUATION OF COMPUTER SYSTEMS FOR
THE WESTERN FARM IMPLEMENT COMPANY

Prepared for:

The Western Farm Implement Company
Foothills, Colorado

Prepared by:

Brian E. Lundeen
Computer Consultant

Report No: E-28   December 19, 19xx

SUMMARY

If it is to continue as a profitable, growing organization, The Western Farm Implement Company should convert to a computerized accounting and inventory control system. This will eliminate the high costs and inefficiency inherent in the present manual system.

Our study shows that the best method for accomplishing this changeover is to lease computer time from Datashare Inc for three years, and to purchase terminals for in-house installation from Vancourt Computer Systems. The cost will average $14,320 per year.

After three years, we recommend that The Western Farm Implement Company purchase its own computer system. We suggest a further study be conducted then to identify the company's requirements.

TABLE OF CONTENTS

	Page
Summary	i
Table of Contents	ii
Introduction	1
Comparison of Manual and Computer Systems	1
Existing Manual System	1
Proposed Computer System	2
Background Information on Computers	3
Hardware	3
Software	4
General Comments	4
Evaluation Considerations	4
In-house Minicomputer Systems	5
Echobyte Computers Model 412	6
Compugraphics System 70	6
Timesharing Computer Systems	7
The Datashare Inc Timesharing Plan	8
In-house Input and Output Terminals	8
Comparative Costs of the Systems	10
Comparison of In-house and Timesharing Systems	10
Conclusions	11
Recommendations	12
References	12

APPENDICES

A - Echobyte Computers Model 412 Minicomputer
B - Compugraphics System 70 Minicomputer
C - Vancourt Computer Systems Input and Output Terminals
D - Comparative Costs of Computer Systems

EVALUATION OF COMPUTER SYSTEMS FOR THE WESTERN FARM IMPLEMENT COMPANY

INTRODUCTION

① This report contains the results of our study into the possible use of a computer to handle the accounting and inventory control functions of The Western Farm Implement Company (WFIC).

As an agricultural implements and parts wholesaler, WFIC has shown a steady increase in business over the last three years. During this period, however, the company's costs for manual accounting and inventory control have risen to $30,000 per year. In a letter dated November 2, 19xx, Ms. Anne D. Martin requested us to determine whether the volume of business warranted using a computer for these tasks and, if so, what type would best suit the company's needs.

We have limited our considerations to the options of installing an in-house minicomputer or buying time (timesharing) on a large external computer. We have excluded microprocessors from our report, as those we examined have neither the versatility nor the storage capabilities to meet the company's projected needs. For each of the above options we have evaluated two systems which we consider are representative of the broad range of systems available.

COMPARISON OF MANUAL AND COMPUTER SYSTEMS

② The decision to convert from a manual to a computer system is important, because a great deal of time, effort, and money are required for a successful changeover. Our examination shows that such a conversion is not only desirable but also essential to the long term success of WFIC.

Existing Manual System

The major problem with the manual system is that it is slow and prone to human error. WFIC raised more than 12,000 invoices and purchase orders in the past fiscal year, all of which had to be typed manually. As even a fast typist can type only 8-10 characters per second, this represents a considerable workload; and further time must then be spent checking the typed work to ensure its accuracy and that information has been entered in the proper spaces on the forms. Subsequently locating a particular piece of information, such as an old invoice or purchase order, requires a time-consuming search through the company's bulk records.

1

As WFIC stocks about 800 items, it is impossible to constantly check that appropriate supply levels are being maintained, or to detect slow-moving items. Consequently, inventory is larger than it has to be to cover the needs of the company's clients. Finally, it is difficult to obtain a general overview of company operations with a manual system, which hinders management's efforts to identify and correct problem areas.

To continue with a manual system as WFIC grows will result in errors, higher costs due to wasted time, larger-than-necessary inventory, and a consequent reduction in profits and growth.

Proposed Computer System

With a properly programmed computer system all accounting and inventory control presently done manually can be done with far greater efficiency in a fraction of the time and at no increase in cost.

The entire records of The Western Farm Implement Company, from invoices to stock levels, can be stored on magnetic disks, reducing information retrieval time to a few seconds. And all data from the computer can be printed automatically and accurately onto forms at over 100 characters per second.

Inventory can be reduced significantly by programs that monitor stock levels and automatically print out a purchase order when stock falls below a pre-selected level. The computer will also carry out all accounting functions, eliminating the possibility of human error from all phases except data entry.

With the wealth of information a computer can provide, management can quickly correct minor problems before they become serious. The result is a smooth, efficient operation which can maximize its profits and growth rate.

The benefits of a computer system have been described by Thelma Forbes[1], who points out that a small business with a manual inventory system can lose up to 10% of its total business volume each year simply because stock levels are improperly predicted and maintained, and records are tardily updated. In comparison, a computer system, which anticipates rather than reacts to a company's needs, can increase business volume and replace that lost 10% each year. However, a computer system can just as easily become a liability if it is too small, too large, overly expensive, or improperly programmed. Plainly, not any system will do.

BACKGROUND INFORMATION ON COMPUTERS

People working in the computer field use terminology which makes computers seem more complex than they really are. To help you evaluate our report, this section describes the "hardware" and "software" of computers, and defines some of the more common terms.

Hardware

In general, all physical components of a system are referred to as hardware. In most cases, this consists of a CRT terminal, the central processing unit (CPU), the disk drives and storage disks, and a line printer.

The CRT terminal is a combination typewriter-style keyboard and TV screen, and is used to enter data into the CPU. Input being entered, and output from the CPU, can be displayed on the CRT's screen or sent to the printer to be typed as "hard copy." Most line printers can type at 100-200 characters per second.

The central processing unit (CPU) is the heart of the system, as it contains the working memory where programs are stored and employed. A memory contains small bits of information, which are usually measured in thousands, with one thousand bits being labelled as one "K." A memory capacity of 32K (32,000 bits) would be sufficient for WFIC's initial needs, and could easily be extended by adding a second memory board at a later date.

A working memory requires a constant supply of power, however, and any programs and data in it will be lost if there is a power failure. To guard against such losses, programs and data are transferred from the working memory and entered onto magnetic storage disks. Disk storage capacity is usually a few hundred K, although some systems offer disks with memories of over 1000K. The disk drive is used to locate and enter information onto the disks. Initially, two drives would be sufficient for your needs, and additional drives can easily be added to most systems.

The flow chart in Figure 1 illustrates how the different components of a computer system interact, with arrows indicating the direction of data flow.

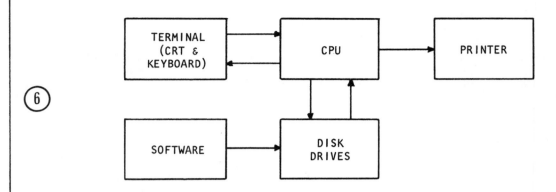

Figure 1: Block diagram of main computer components

Software

The term software refers to the programs that the computer uses for its accounting and inventory control functions. These programs can be purchased as part of a complete system or separately from software vendors. Custom programs can also be designed to meet a particular company's needs, but are considerably more expensive than standard programs. These programs are stored on disks and are transferred into the CPU memory whenever they need to be used.

General Comments

Since the computer field is highly competitive, computer reliability is generally very good. However, even the best systems occasionally experience outages. For this reason we believe that the availability of a local service organization, able to provide both periodic preventive maintenance and to respond to service calls at short notice, is an important consideration. We also consider it wise to purchase from an older, established company with a proven reliability and service record.

EVALUATION CONSIDERATIONS

In evaluating the various systems we considered a number of factors. Foremost, the system had to meet most of the capabilities desired by WFIC. From our discussions with Ms. Martin we determined that the chosen system must be able to handle the following functions:

1. Inventory control, which consists of:

    * Continuous identification of stock levels.
    * Automatic production of a machine printed purchase order when the stock of a particular item drops to a predetermined level.
    * Detection of slow moving items.

2. General accounting, which consists of:

    * Accounts payable.
    * Accounts receivable.
    * Profit and loss statements.
    * Taxation data.
    * Payroll records.
    * Automatic check issuing.
    * Address lists of clients and suppliers.
    * Form type correspondence.

Early in our study we decided to omit payroll records as a necessary function. Both system users and computer companies agree that it is uneconomical for a company with under 50 employees to buy a payroll journal program for their computer system[2]. WFIC has only 23 employees.

Subsequently, we decided to eliminate the ability to produce form-type correspondence. Although at first it seemed a "nice to have" feature, from discussions with system users we discovered that unless a company produces a lot of form-type correspondence such a feature becomes an expensive luxury. Consequently, none of the systems we evaluated included this function.

Other factors we considered were system cost, storage capability, expandability, adaptability, servicing costs, and operator training time.

IN-HOUSE MINICOMPUTER SYSTEMS

The term in-house refers to the fact that all hardware and software are located on the company's premises. The two systems we evaluated were the Echobyte 412 and the Compugraphics System 70.

⑫ Echobyte Computers Model 412

⑬ A basic starter system capable of satisfying WFIC's present needs could be purchased for only $17,200. The Echobyte 412 is typical of the single-unit computers currently available. It includes a CRT terminal, CPU with 32K memory, a single disk drive for 360K capacity disks, and a printer, all housed in one cabinet. However, the manufacturer does not expand the capability of its preassembled units after initial purchase, which means that any future expansion requirements must be built into the system as part of the original order. Thus, for a system expandable to 48K CPU capacity, with two disk drives, the purchase cost would increase to about $18,500.

⑭

⑮ The low cost of this system is offset by two factors. First, the printer, while providing high quality hard copy, has a slow print rate (for a computer printer) of only 55 characters per second. Secondly, all quoted costs are for hardware only: programs for the system must be purchased separately. The 412 is capable of using either user-programmable packages which Echobyte will design for its clients, or pre-programmed packages which can be purchased from independent software vendors. Either will provide a highly adaptable system capable of handling all the desired functions, but we consider both options to be inconvenient and possibly very expensive. For example, a minimum set of pre-programmed packages would cost at least $10,000. User-programmable packages would cost twice as much.

A major disadvantage to this system is remote servicing. Like many relatively small manufacturers, Echobyte has no local sales or service center, and until it does all warranty repairs must be made at the manufacturer's facilities in Connecticut. Neither does the Echobyte literature identify how much operator training is required. The local sales representative (Jim Reiter of Dual Electronics) could not answer our questions on maintenance and training.

⑯ Additional information and specifications on the Echobyte 412 are contained in Appendix A.

Compugraphics System 70

⑰ The Compugraphics System 70 is one of several minicomputer systems which are sold as complete packages, including all necessary hardware and software. It costs $34,000.

⑱ The hardware comprises the CRT input terminal, the central processor with two disk drives, and a line printer capable of producing high quality hard copy at 150 characters per second. The disk storage capability is extremely large, rated at 1200K for the two-drive standard system. The system can

6

be expanded even further, to a total capacity of 19,000K, and can be switched to a multi-terminal system by changing one circuit board in the central processing unit (CPU).

(19) Its software package can handle all the functions required by WFIC (excluding form-type correspondence) and more. A detailed list of its pre-programmed functions, plus information on the system, is contained in Appendix B. Since these programs come as a standard, pre-designed package, you would likely have to modify your accounting and inventory control methods to them, rather than try to adapt them to your present system. This would not be difficult to accomplish.

(20) On the negative side, like Echobyte, Compugraphics does not operate its own local service facility. It has appointed MicroPro Systems to service its equipment during the warranty period. As MicroPro Systems has been in the computer sales and service business for only 18 months, their reliability and service record have not been firmly established.

(21) The brochure describing the System 70 claims that operators can be trained to use the system in half a day. We suggest 1½ or 2 days would be a more likely estimate for this familiarization period.

TIMESHARING COMPUTER SYSTEMS

(22) In a timesharing system, only the input and output terminals, consisting of keyboards, CRT screens, and line printers, are located in the user's building. The central processing unit, disks, and disk drives are all part of a large computer owned by the timesharing company. The timesharing company this report examines is Datashare Inc.

Clients leasing computer time are connected to the timesharing company computer by telephone. The telephone connection can be made either over normal telephone lines or by a private line especially installed by Foothills Telephone Company and used solely for data transmission. Such a connection is known as a dedicated line, and it becomes economical when a client uses the computer for more than six to eight hours per week. There is no extra charge for a normal telephone connection. A dedicated line costs a basic $60.00 for initial installation, followed by a monthly rental of $5.00 per mile. As 5.6 miles of line would be needed to connect WFIC to Datashare Inc, the rental cost would be $28.00 per month.

7

The Datashare Inc Timesharing Plan

The cost of leasing computer time from Datashare is $8.00 per hour with a regular telephone line, or a flat rate of $250 per month with a dedicated line. The dedicated line becomes cheaper when the computer is used more than 100 minutes each day (of a five-day week), as we predict it would be by WFIC. There is no problem with storage capabilities or future expansion because the computer is a full-size mainframe, much larger than any mini-computer available.

All data is backed up (put on permanent storage disks) twice daily, ensuring that only half a day's data would be lost if a power failure occurs. Historical records we examined show that Datashare has lost a client's data only twice during the past eight years, and on both occasions during system upgrading. The company has never lost data from other causes. As a further protection, Datashare insures all data against such mishaps. Down time, or time during which the computer is not available for use because of servicing or an outage, is about 0.1%, or less than 10 hours per year.

Datashare offers a complete line of prepared programs (software) which would be suitable for WFIC's requirements. Lease costs for the minimum software needed would amount to $320 per month. Program modifications and adaptions to meet your specific needs would cost no more than $600.

Servicing of the processor and storage systems is handled entirely by Datashare, with the cost being covered by the hourly or monthly rental fee. From the user's viewpoint this is a major advantage compared to an in-house system, since maintenance problems are most likely to occur in the CPU and disk drives.

A half-day system familiarization session and operator training are included in the price of the Datashare contract, as are periodic checks to ensure smooth operation of the system. Your existing staff could be trained to operate the terminals, since timesharing uses a "BASIC" computer language. Datashare states that it can train a client's operators in half a day; we believe one day would be a more realistic figure.

In-house Input and Output Terminals

Because Datashare does not provide CRT terminals and printers, they must be either leased or purchased from a separate supplier. In general, purchasing is more economical if the terminals are to be used for more than two years. Purchased equipment also is eligible for investment tax credits, amortization, and special tax treatment for depreciation[3]. With the purchase, however, comes the responsibility for equipment servicing.

8

We have examined three representative sources from which terminals may be leased or purchased:

1. Foothills Telephone Company

    Like most telephone utilities, Foothills Telephone Company not only sells telephone services but also offers a variety of input and output terminals on a leasing arrangement. CRT terminals are available for $115 per month and line printers for $132 per month. There is a combined initial installation fee of $125. The total cost for one terminal and one line printer amounts to slightly less than $3100 per year.

2. Echobyte Computers

    In addition to their complete system, Echobyte Computers sells separate components such as the T2040 CRT terminal and the P1850 line printer. Both are part of the System 412 described earlier, and their specifications are shown in Appendix A. The CRT sells for $995 and the printer for $2915, giving a total cost of $3910. As mentioned earlier, however, the printer has a very slow print rate, and both it and the CRT must be serviced in Connecticut.

3. Vancourt Computer Systems

    Components sold by Vancourt Computer Systems which would best meet your needs are their Vudat CRT terminal and Dataprint line printer. At $1295, the Vudat is $300 more expensive than the Echobyte T2040; and the Dataprint sells for $2955, comparable to the price of the Echobyte P1850. The combined cost of the Vancourt terminal and printer would be $4250.

    The Vancourt terminals have the advantage that they can be serviced by the local branch of Vancourt Business Systems Inc. Maintenance and service contracts can be purchased at $17 per month for the CRT terminal and $25 per month for the printer, for a combined cost of $42 per month or $505 per year.

    The Dataprint also offers a high speed print rate of 180 characters per second and has its own keyboard, allowing dual access to the system. Further information and specifications on these two terminals are included in Appendix C.

9

COMPARATIVE COSTS OF THE SYSTEMS

A more detailed treatment of this topic is given in Appendix D. For simplicity, only total costs are presented here.

Of the two in-house systems, the Echobyte 412 is less expensive. It would cost about $32,200 the first year, and $3700 per year after that, for a total cost of $39,600 over three years. This low price, however, is only for hardware; software would cost at least $10,000 more, bringing the total cost up to a minimum $50,000.

The Compugraphics System 70 is the more expensive in-house system, costing $52,500 the first year and $8500 yearly for the second and third years. This gives a three-year investment of almost $70,000, but includes the cost of software.

The Datashare timesharing system, with terminals leased from Foothills Telephone Company, would require a first year outlay of almost $21,950, and about $12,275 per year for the next two years. Over three years, this system would cost approximately $46,500.

For the same timesharing plan, but with terminals purchased from Vancourt Computers Inc, the cost would be $23,591 the first year, but only $9681 for the second and third years. This give a three-year expenditure of $42,953.

COMPARISON OF IN-HOUSE AND TIMESHARING SYSTEMS

A purchased in-house system has several advantages over leasing time on a large computer. Purchased equipment is eligible for tax credits and has a lower yearly cost after the initial investment has been made. However, with a purchased system WFIC would have to pay the costs of all servicing, system expansion, and system and data protection. Should major repairs become necessary, the cost could become a heavy financial burden on the company.

A timesharing system eliminates many of these concerns, as the mainframe computer company (Datashare Inc) would be responsible for maintaining the most costly components (the CPU, disk drives, and storage disks). This transfer of responsibility is paid for in the average higher yearly cost of timesharing compared to complete ownership. For this reason timesharing is less economical over a long period of time; eventually a point is reached when a purchased system becomes more economical.

A point which must be considered here is the annual gross income of WFIC. As a rule of thumb, the computer industry has established $2.7 million as the average minimum annual gross income for any company considering purchasing its own computer system[3]. At its present growth rate, WFIC should reach this level in four or five years. However, with an improved operation which should result from using a computer, this level could be reached earlier, possibly in only three years.

A timesharing system operated over this three-year period would cost as little as $42,950, or about $14,320 a year. After this initial three years, WFIC should be in a stronger financial position to consider the purchase of an in-house minicomputer system. A second study would establish the viability and timing of such a conversion.

CONCLUSIONS

To increase its efficiency and profits, The Western Farm Implement Company should change over to a computerized accounting and inventory control system.

WFIC has the alternative of purchasing a complete in-house minicomputer, or leasing time (timesharing) on a large remote computer. The cheapest in-house minicomputer would cost a minimum $32,200 the first year, and $3700 for each subsequent year, for a total three-year price of $39,600 (plus the cost of software). The cheapest timesharing system would cost about $23,590 the first year, and $9680 for each subsequent year, for a three-year price of $42,950, including software.

The low start-up cost and "worry-free" aspects of timesharing make it a more attractive initial system, particularly for companies with a gross annual income of less than $2.7 million. In three to four years -- when WFIC's annual income is predicted to surpass this figure -- it will probably be more economical and efficient for WFIC to purchase its own computer.

11

RECOMMENDATIONS

We recommend that The Western Farm Implement Company convert from a manual to a computerized accounting and inventory control system.

To implement this changeover we recommend:

1. Renting computer time from Datashare Inc for three years.

2. Leasing a dedicated line from Foothills Telephone Company, to connect WFIC with Datashare's computer.

3. Purchasing input and output terminals from Vancourt Computer Systems; specifically, the Vudat CRT terminal and Dataprint line printer.

REFERENCES

1. Thelma Forbes, "Even Small Businesses Can Benefit from a Computer" in Independent Business Managers Journal, 7:06, June 1981, p. 62.

2. Allan B. Shaw, "Pros and Cons of Computerized Payrolls" in Making Tracks in Data Communications, January 26, 1979, p. 143.

3. How to Select and Install a Small Business Computer, Report No. E80 300 401, Datapro Research Corporation, Delran, NJ, April 1977.

APPENDIX D

COMPARATIVE COSTS OF COMPUTER SYSTEMS

The prices and costs quoted in this appendix are not exact, because some assumptions and generalizations had to be made to arrive at the figures quoted on the next page. For comparison purposes, however, they are sufficiently accurate.

The maintenance cost quoted is based on a general rule of thumb obtained from conversations with computer consultants, and on research undertaken by Datapro Research Corporation. As a rule, maintenance costs over five years will equal the initial cost of the hardware, or one fifth of the purchase cost per year.

A study by Datapro[1] estimates start up costs to be around $10,000. We have reduced this figure to $9000 for a timesharing system because the computer is already set up and functioning. We have also estimated a yearly cost of $2000 for supplies such as paper, forms, and printer ribbons.

An approximate yearly cost breakdown for each system is shown in Table 1.

---

[1] *Small Business Computers -- A Market Overview*, Report No. E30 160 101, Datapro Research Corporation, Delran, NJ, January 1979.

Appendix D

TABLE 1 - APPROXIMATE YEARLY COST BREAKDOWN FOR ALTERNATIVE COMPUTERS

System		1st Year	2nd Year	3rd Year	Three-year Total
Minicomputer Systems	Echobyte Computers Model 412	Hardware (computer) $18,500 Start-up costs 10,000 Maintenance 1,700 Supplies 2,000 TOTAL $32,200	$ -- -- 1,700 2,000 $ 3,700	$ -- -- 1,700 2,000 $ 3,700	$39,600 (excluding software) $49,600 (including software)
	Compugraphics System 70	Hardware (computer) $34,000 Start-up costs 10,000 Maintenance 6,500 Supplies 2,000 TOTAL $52,500	$ -- -- 6,500 2,000 $ 8,500	$ -- -- 6,500 2,000 $ 8,500	$69,500 (including software)
Timesharing Options	Datashare Computer with Leased Terminals (Foothills Tel. Co)	Computer time $ 3,000 Software 4,440 Terminals (leased) 3,100 Dedicated Tel. line 396 Start-up costs 9,000 Maintenance -- Supplies 2,000 TOTAL $21,936	$ 3,000 3,840 3,100 336 -- 2,000 $12,276	$ 3,000 3,840 3,100 336 -- 2,000 $12,276	$46,488
	Datashare Computer with Purchased Terminals (Vancourt)	Computer time $ 3,000 Software 4,440 Terminals (purchased) 4,250 Dedicated Tel. line 396 Start-up costs 9,000 Maintenance 505 Supplies 2,000 TOTAL $23,591	$ 3,000 3,840 -- 336 -- 505 2,000 $ 9,681	$ 3,000 3,840 -- 336 -- 505 2,000 $ 9,681	$42,953

V
REPORT WRITING TECHNIQUES AND METHODS

9
The Appearance and Format of Letter and Memorandum Reports

The appearance of the reports you write should echo the quality of your words. A poorly presented report can create the impression that what you have written is poorly expressed; or, worse, that the information you are presenting is of secondary importance. Readers will gain this impression as soon as they pick up your report—before they read a word—and their reaction to the information you present may be subtly downgraded.

On the other hand, a neatly presented report, in the proper format to suit the circumstances and the intended reader, will create the impression that you are presenting valuable information. In the reader's eyes it will enhance your credibility as a reporter of information, again before he or she starts reading. It can subtly assist acceptance of your facts and figures.

This chapter will help you choose the proper shape for each report you write. There are four formats to choose from:

1. A memorandum, which is used when a report is directed from one person to another within the same organization. It is an informal form of presentation.
2. A letter, which is normally used when the writer belongs to one organization and the person to whom the report is directed (to whom the letter is addressed) belongs to another organization. Letters are more formal than memorandums, but are still an informal reporting medium.
3. A titled document, in which the report's title and the author's name are centered at the head of the first page, with the report narrative starting beneath them. Because its appearance is slightly more formal than that of the letter, it is referred to as a semiformal report.

134 *The Appearance and Format of Letter and Memorandum Reports*

   4. A bound document, with a cover and full title page preceding the report proper, and separate pages for individual sections such as the summary and table of contents. Such reports are known as formal reports.

Guidelines for presenting informal and semiformal reports in the correct format are outlined in the sample memorandum, letters, and first page of a semiformal report illustrated in Figures 9-1 through 9-4 of this chapter. Guidelines for presenting a formal report are included with the report analysis in Chapter 8.

Legend for Figures 9-2 and 9-3 (Pp. 136–37)

A. It is customary to name the person to whom a letter is addressed first, and to follow the person's name with his or her title and then the name of the company or organization. The position the person holds may be placed in the second line (as in Fig. 9-2) or beside the person's name (Fig. 9-3).

B. Punctuation is omitted from the recipient's address, except where a comma is needed to separate two unrelated words in the same line. Although it is more common to insert a period after "Mr.," "Ms.," and a person's initials, there is a trend to omit such punctuation.

C. The use of a colon (:) after the salutation and a comma (,) after the complimentary close is optional, but their insertion or deletion should be consistent.

D. Subject lines are optional. If used they should be underlined, and may be preceded by "Subject:," "Ref:," or "Re:" (also optional).

H. L. Winman and Associates

INTER - OFFICE MEMORANDUM

From: R Bryant
To: W K Carter
cc L L Sampson

Date: June 26, 19xx
Subject: Format for memorandum reports

Informality should be the keyword for your memorandum reports, both in appearance and language. The appearance should be simple, with the salutation and signature block omitted.

Informality, however, should not be interpreted as a signal that you can use sloppy language. Your reports must be organized coherently; your paragraphs and sentences must be properly constructed; and the words you use must be simple and clear.

You may use subparagraphs (point form) to break up long paragraphs or to list a series of points, as shown below. And if a report is long you may insert headings to help readers see how you have organized your information. Other factors you should be aware of are:

1. The subject line should be _informative_, so that it gives a clear indication of the memo's contents. "Memorandum reports" would not have been sufficiently informative as a subject line for this memo.

2. The first line of each paragraph may be indented five spaces or started flush with the left-hand margin.

3. The names of persons who are to receive carbon copies of the memo may appear directly under the addressee's name (as has been done at the top of this memo) or entered three lines below the last paragraph, against the left-hand margin.

You may either sign or initial a memorandum report.

Rod Bryant

Fig. 9-1. The shape of an informal memorandum report.

VANCOURT BUSINESS SYSTEMS INC

September 27, 19xx

(A) Mavis J. Morgan
Manager, Customer Services
Friesen Discount Sales Inc
(B) 2820 Border Road
Sienna CO 81506

(C) Dear Ms. Morgan:

(D) <u>Letter Reports in Modified Block Format</u>

The modified block format is the more conservative of the two letter styles currently used by business and industry. Because it does not appear to be as severely businesslike as the full block format, it is particularly suitable for writing to the public.

Both the left-hand margin and the page centerline are used to position the various parts of the letter horizontally. For example:

1. The name and address of the person you are writing to are typed flush with the left-hand margin ("flush" means all lines start at the margin).

2. The first line of each paragraph may start at the margin or be indented five characters (letter spaces), as has been done here.

3. The date and the signature block start at the page centerline.

4. The subject line is centered on the page centerline, and normally is underlined.

5. The left- and right-hand margins are roughly the same width.

If the report is short enough to fit on one page, it should be positioned vertically so that the body of the letter is in the middle of the page.

(C)   Sincerely,

Franklyn Weatherdon

Franklyn T. Weatherdon
Publications Editor

FTW:ma

Fig. 9-2. The shape of a letter report in modified block format.

March 18, 19xx

THE RONING GROUP

Communication Consultants

File: 276-13

(A) Mr. R. Craig Williams, Head
Corporate Resources Division
Centaur Corporation
(B) P O Box 2760
Baton Rouge, LA 71036

(C) Dear Craig

(D) The "Full Block" Letter Format

My analysis of 300 major companies in the U.S. and Canada shows that 246, or 82%, prefer the full block letter style for their corporate correspondence and informal letter reports.

In the full block format every line starts at the left-hand margin, which simplifies typing because no lines have to be measured and positioned about the page centerline. However, shorter (one page) letter reports have to be carefully centered vertically on the page if they are to achieve a balanced appearance.

Because it conveys the impression of a modern, forward-thinking organization, I recommend you adopt the full block style for the Centaur Corporation's correspondence.

(C) Regards

Marilyn Duvall

Marilyn Duvall, Specialist
Business Communications

MD:es

Fig. 9-3. The shape of a letter report in full block format.

Legend for Figure 9-4

1. The report title should be positioned about 1½ to 2 inches below the normal top line of typing (i.e. about 3 to 3½ inches below the top edge of the page), to make the first page appear better balanced and less forbidding.

2. The author's name and affiliation may appear either here or on the last page of the narrative (i.e. ahead of the attachments).

3. The first line of each paragraph may be indented five spaces or may start at the left-hand margin, as in Figure 9-4. The latter method is preferred.

4. To create an uncrowded appearance there should be 1½ or 2 blank lines between paragraphs, but only one blank line between a heading and the paragraph that follows it. See Lorraine Dychuk's proposal on pages 86–93 for an example.

---

① PREFERRED FORMAT FOR SEMIFORMAL REPORTS

② Rodney T. Elson
Communications Consultant

③ Summary

The appearance of a semiformal report lies half way between the comfortable informality of the letter report and the strict formality of the formal report. The report's title is given prominence by being displayed in capital letters across the upper center of the first page, and the author's name and company affiliation are centered neatly beneath it. The narrative of the report follows immediately, and continues onto subsequent pages. The impression gained be a reader on first seeing the report should be of a quality document containing important information.

④ The Report's Parts

The parts of a semiformal report are similar to those of a formal report, except that it normally has no cover, the summary seldom has a page to itself, and the table of contents page is omitted. As in a formal report, each major section is introduced by a center or side heading.

Fig. 9-4. The appearance of a semiformal report (top of first page only).

10
The Language of Report Writing

We use essentially the same language for report writing as we do for regular correspondence and day-to-day writing tasks: the primary difference is that report writers must strive particularly to be brief yet fully informative. A report must contain all the information its readers will need to thoroughly understand a given situation and take any necessary action, and yet it must neither waste the readers' time by conveying too many details nor obscure the message by using ponderous sentences and paragraphs. This chapter describes techniques which will help you focus your reports correctly, be direct, and avoid cluttering the narrative with unnecessary words and expressions.

GET THE FOCUS RIGHT

The first rule of report writing is to remember never to start writing until you have answered three questions. They are:

1. Who is my reader?
2. What is the purpose of my report?
3. Do I want to be purely informative or convincingly persuasive?

Your answers will give you a sense of direction, and help you write much more easily and spontaneously than if you had simply picked up a pen and started writing. The implications of each question are outlined below.

Identify the Reader

Karen Williams has been studying materials handling methods used by her company and reckons there is a better way to manage the ordering, receiving, documenting, storing, and issuing of parts and materials. Her department manager has told Karen that she is to write a report of her findings and recommendations, but has omitted to tell her who will be reading the report and using the information it contains.

Karen is likely to make many false starts if she starts writing her report without having a clearly defined audience in mind. She may feel she is writing in a vacuum because she cannot direct her words to a specific person or to a group of people. Without a known target, she will not be able to focus her report.

Step number one in any writing situation is to identify a specific person as your primary reader, and a general group of people who may also see your report as your secondary readers. If you are not told who your readers are, and how they will use your report, you may have to ask.

Karen Williams cannot assume she is writing a report only for her department manager's eyes. If her manager simply wants information which he can use as the basis for a proposal he plans to write and submit to the head office, then Karen's report is for only one reader. But if he plans to use Karen's report as evidence which he can submit to the head office under a covering memorandum, then it will be read by a much wider audience. Hence, to focus her report properly she needs to know clearly who will read it and how it will be used.

Identify the Purpose

Always check that you have clearly identified the purpose of your report before you start writing. On a separate sheet write the words:
   "The purpose of my report is . . ."
and complete the sentence (but limit yourself to only one sentence). Karen Williams, for example, should write something like this:

> The purpose of my report is to demonstrate that our materials handling methods are outdated, and to show how they can be improved.

Now Karen knows both for whom and why she is writing. Yet she still has to decide whether her report is to be informative or persuasive (i.e. if it is to "tell" or to "sell").

When a report writer is simply informing readers of a given situation, such as progress in compiling a client index, and does not have to persuade readers to take any action as a result of the report, then the purpose is solely to present facts. But if a report writer wants to evoke a response from readers, or to convince them to take or approve some action such as hiring an additional person to enable completion of a project by a certain deadline, then the purpose is to convince the readers of the validity of the case and to persuade them to act. This can be difficult if the required action will cost money or demand a change in established routines.

If Karen has identified that her department manager is to be the only reader, and she knows he is seeking only facts, then she will write an informative report because she simply has to tell him what she has found out. Alternatively, if she has identified that her report will be sent to the head office to convince executives to invest in a new materials-handling system, she will have to write a persuasive report because she also has to convince management that the new system is necessary.

Write to Inform

Informative writing is much simpler than persuasive writing. To write informatively you need to present facts clearly and in logical sequence. You should write briefly, directly, and forthrightly, closely following the pyramid structure described in Chapters 2, 3, and 4. The pricing error report in Chapter 3, and the mobile trailer progress report in Chapter 4, are typical examples of informative writing.

Write to Persuade

The difficulty with persuasive writing is preserving one's objectivity. Although your aim should be to convince readers to accept your ideas, your partiality or bias should not be so obvious that readers feel they are being coerced. Hence, how you write your report is extremely important if you do not want to turn your readers off.

Fortunately, several sections of every persuasive report deal with facts, and these can still be presented informatively. These are mainly the introductory section, in which you describe the background of your report, and the description of your approach and findings. (Karen's description of the existing materials handling methods, for example, should be strictly informative regardless of whether she is writing a "tell" or "sell" report.) Only when you have to present your suggestions and analyze advantages and disadvantages should your involvement become obvious. Of course, when you make a recommendation your preference should be readily apparent.

The suggestion and proposal in Chapter 7 and the formal report in Chapter 8 are examples of persuasive writing. The two longer reports, particularly, show how their authors have gradually developed their cases, working carefully from an informative presentation of facts toward a persuasive evaluation of alternatives.

BE BRIEF AND DIRECT

Chapters 2 through 8 stress the need to satisfy a reader's curiosity by identifying the most important information, consolidating it into a short summary statement, and placing it right at the front of every report you write. This "direct" writing technique can be extended to individual sections of a report and to each paragraph. It can also be enhanced by writing as much as possible in the first person and in the active voice.

Use the Pyramid Structure

There is no need to feel that the pyramid structure described in Chapter 2 applies only to a complete document. Within a long report each major section also should be structured "pyramid style," with each section opening with a short summary statement followed by the basic BACKGROUND-FACTS-OUTCOME arrangement of information. This technique is even used in this chapter, with the opening paragraph of each major subsection starting with a summary statement. For example, the side heading GET THE FOCUS RIGHT on page 139 is followed immediately by the short paragraph:

> The first rule of report writing is to remember never to start writing until you have answered three questions. They are:
> 1. Who is my reader?
> 2. What is the purpose of my report?
> 3. Do I want to be purely informative or convincingly persuasive?

This paragraph identifies the topics discussed in the remainder of the subsection.

Similarly, the first paragraph of this subsection (immediately following the heading BE DIRECT) summarizes what is being described here and on the next two pages.

In turn, each paragraph can also be structured "pyramid style," with the first sentence being a topic sentence (summary statement), and the remaining sentences amplifying and developing the initial statement. For example, in the two paragraphs that follow, the topic sentences (in italics) each describe the main point while the remaining sentences provide more details:

1. *We have evaluated the condition of the Merrywell Building and find it to be structurally sound.* The underpinning done in 1948 by the previous owner was completely successful and there are still no cracks or signs of further settling. Some additional shoring will be required at the head of the elevator shaft immediately above the 9th floor, but this will be routine work that the elevator manufacturer would expect to do in an old building.[1]
2. *Many writers find it difficult to pay compliments or to apologize sincerely.* Frequently they set an unnatural tone by trying to say too much. A simple, brief statement seems incomplete, so they "beef it up a bit" under the false impression that they are showing their true feelings. The resulting tone is so false, and the words are so forced, that the reader immediately senses that the compliment or apology is insincere. Such writers have to learn that sincerity is enhanced by brevity: pay your compliment or make your apology, then forget about it.[2]

In high school and college or university you were probably told that every paragraph must have a topic sentence, and that it can be placed:

- as the first sentence,
- as the last sentence,
- as both the first and last sentences,

[1] Ron S. Blicq, *Technically-Write!*, 2nd ed. (Englewood Cliffs, NJ: Prentice-Hall, Inc., 1981), p. 198.
[2] Blicq, p. 20.

or it can even be simply implied. For report writing—and particularly for short reports—you would be wise to place your topic sentence at the beginning of each paragraph. This way your readers can quickly discover what each paragraph is about and so will more readily comprehend the details which follow.

As your experience as a report writer grows, you will undoubtedly learn to use nonpyramid paragraphs successfully in longer reports. But always try to remember that for short, informative reports the pyramid-style paragraph (i.e. with a topic sentence at the beginning) is more direct and more effective than the nonpyramid paragraph. Nonpyramid paragraphs are generally more suitable for the longer persuasive reports, but even then they need to be used only occasionally and to create a particular effect. For example, the events described in the following paragraph lead up to the main point the author wants to make (in the italicized topic sentence):

> We first monitored sound levels between 8 p.m. and 10 p.m., to establish a background sound level while the building was empty. Then the following day we measured sound levels hourly from 7 a.m. to 6 p.m. at 28 locations throughout the building, and recorded the results in Appendix B. Seventeen of the measuring points were in or adjacent to departments where employees had complained of excessively high noise, and eleven were control points in departments from which no complaints had been received. *Although the sound levels measured at the "noisy" locations were an average 7 decibels higher than at the control locations, at no location was the sound level higher than 63 decibels.*

This climactic method (leading up to the main point) can be useful if you have to prove your case to readers who are prejudiced against or tend to resist the facts you have to present. But the same information can just as easily be presented "pyramid style" to readers who are more ready to accept the facts:

> *We monitored sound levels and found that they did not exceed 63 decibels anywhere in the building, although the sound level for departments reporting excessively high noise was an average 7 decibels higher than the sound level for other departments.* Our measurements were recorded on two separate occasions: between 8 p.m. and 10 p.m. one evening to establish a background sound level while the building was empty, and hourly from 7 a.m. to 6 p.m. the following day. Seventeen of the 28 measuring points were in or adjacent to departments where employees had registered complaints, and eleven were control points in departments from which no complaints had been received. The results are shown in Appendix B.

Which approach you use will depend on the effect you want to create. Hopefully, you will want to write pyramid-style paragraphs more often than climactic paragraphs.

Write in the First Person

Reports are written and read by people, so it should be natural to write from person to person. Yet many report writers try to avoid using the first person when they write because they feel they are being "unbusinesslike" or "unprofessional." They write:

> The components have been ordered...
> A data survey was conducted...
> A report was submitted...
> It is recommended that...

The writers of these statements seem afraid to display their involvement in the order for components, the data survey, the report submission, and the recommendation. Perhaps they write in this indirect way because in high school or college English classes they were told it is too "pushy" to use "I" and only royalty and newspaper editors use "we." If so, they have been conditioned to write in an impersonal, indirect way which is forcing them to seem indifferent in their readers' eyes.

Their statements would be much more direct and effective if a "person" could be inserted into them:

> **I** have ordered the components...
> **We** have conducted a data survey...
> **I** submitted my report...
> **I** recommend that...

To write in the first person (i.e., to use "I," "we," "me," and "my") is not unprofessional or unbusinesslike. When you write a report to your manager, to someone in another department, or to someone outside your own organization, you should try to write from person to person. Insert "I" if you are writing from yourself, and "we" if you are reporting for a group of people, your department, or your company or organization.

If you are drafting a report for another person's signature (your department head, for example), you may feel you do not have the right to use "I" and "we." Under these circumstances you should go to the person whose name will appear on the report and ask if you can use the first person. You may have to explain what you mean by "the first person," but the few minutes you spend doing so can be an investment for the future: the next time you have to prepare a report for that person's signature you probably will not have to ask, or at least your question can be briefly posed and answered.

There are numerous examples of reports written in the first person throughout Chapters 3 through 8. The pronoun "I" is readily evident in the informal memo reports in Chapters 3 and 4 (particularly see Frank Crane's trip report, Marjorie Franckel's inspection report, and Tom Westholm's investigation report). Only in Paul Thorvaldson's slightly more formal inspection report and Roger Korolick's progress report is "I" less evident: Paul has limited its use primarily to his suggestions (his "outcome" compartment), and Roger uses "I" only when he refers to his concerns and plans.

In the longer reports, the first person singular ("I") is used by Lorraine Dychuk in her in-house proposal in Chapter 7. But both Tod Phillips and Brian Lundeen have preferred to use the plural "we" for their client-directed investigation report (Chapter 6) and formal evaluation report (Chapter 8). Undoubtedly they both felt that, although they had individually conducted their studies, they were reporting for their companies rather than for themselves.

Use the Active Voice

Which would you prefer to write?

A. Carl Dunstan investigated the problem.

or

B. The problem was investigated by Carl Dunstan.

In report writing you should try to be as direct and as brief as possible without losing any information. As both sentences contain the same information, but B is longer and less direct, you should choose sentence A.

Sentence A is written in the active voice, in which the person or object performing the action is stated first; like this:

Carl investigated the problem.
The shaft penetrated the casing.
Petra is studying the charts.

Sentence B is written in the passive voice, in which the person or object performing the action is stated *after* the verb; for example:

The problem was investigated *by Carl*.
The casing was penetrated *by the shaft*.
The charts are being studied *by Petra*.

Sentences written in the active voice are generally shorter and more emphatic than sentences written in the passive voice. Similarly, reports written primarily in the active voice seem much stronger, more definite, and more convincing than reports written predominantly in the passive voice. Compare these two paragraphs, both describing the same situation.

Primarily Passive Voice

A study of electricity costs was conducted in three stages over a twelve-month period. First, a survey was taken and a list made of all apartment dwellers in the area. Then a table was constructed in which family size was compared against apartment size. Finally, an analysis was made of apartment dwellers' life-styles and their major appliance ownership (it was assumed that a stove, refrigerator, and air conditioner were installed as standard equipment in each apartment). *(75 unassertive words)*

Primarily Active Voice

We studied electricity costs in three stages over a twelve-month period. First we surveyed and listed all apartment dwellers in the area, and then constructed a chart comparing family size against apartment size. Finally we analyzed apartment dwellers' life-styles and their major appliance ownership (we assumed that each apartment was equipped with a stove, refrigerator, and air conditioner as standard equipment). *(61 confident words)*

The information conveyed by the two paragraphs is essentially the same, yet the impact each creates is markedly different. The active-voice paragraph seems to be written by a confident, knowledgeable individual who uses the first person ("we") and clearly identifies that someone has been actively doing something. The passive-voice paragraph seems to be written by someone who is detached and uninvolved; its author does not write in the first person, and so writes sentences without mentioning who performed the action. This creates the impression that he or she is merely passing along information.

Writing in the first person can help you avoid writing in the passive voice. When you write "I" or "we," you immediately identify who was involved:

> *I* requested approval to visit . . .
> Early in May *we* established criteria . . .

This becomes particularly important when you have to make recommendations. The Recommendations section of a report must be strong and definite; yet often report writers adopt a more indefinite, passive stance, writing

> It is recommended that . . .

Instead, they should be firm and assertive, and write in the active voice:

> *I recommend* . . .   (when the report writer is making a recommendation as an individual) or
> *We recommend* . . . (when he or she is making a recommendation on behalf of a group of people, the department, or the company)

Tod Phillips and Brian Lundeen both write "We recommend . . ." in the Recommendations sections of their reports (see Chapters 6 and 8), while Lorraine Dychuk writes "I recommend . . ." in her proposal in Chapter 7.

Writing in the active voice does not mean you always have to write in the first person. You can just as easily name another person, a department, or an object:

> *Mr. Swanson* revised the estimate.
> To meet the mailing deadline *the project group* worked until 3 a.m.
> *My bank* increased the interest rate.
> On the fourth floor *the mail cart* lost a wheel.
> Our *comptroller* recommended a budget cut.
> After reading our report, *the client* requested a revision.

Of course, when you do not know who performed the action, prefer not to name names, or want to deemphasize the doer, then the passive voice has to be used:

> Your budget has been cut by 30%. (*Specific person not stated*)
> The documents were misfiled. (*Person not known*)
> The long-awaited Manston report has been printed. (*The emphasis would be wrong in the active voice:* "The printer has printed the long-awaited Manston report.")

This section only draws your attention to the active voice and suggests that you use it wherever possible in your reports. For a more detailed description refer to a language textbook such as the *Prentice-Hall Handbook for Writers*.[1]

AVOID CLUTTER

The words you use in a report can do much to help your readers quickly understand what you have to say, and then react or respond in the way you want them to. A clear, concisely worded report will evoke the correct reader response. But a report cluttered with unnecessary words and expressions can so muffle the message that readers either miss the point or lose interest and stop reading. "Clutter" words are dangerous because they can create an image of you as a very wordy individual who uses numerous cliches and overworked expressions. The problem with clutter words is that we may see them used frequently by *other* people, and so tend to recognize them as old friends; hence we may have difficulty in weeding them out of *our* writing.

Use Simple Words

When you have the choice between two or more words, try using the simpler word. The accountant who refers to "remuneration" and "superannuation scheme" in an annual report would do better to write about pay, salary, or wages, and the pension plan. Then he would be understood by virtually every reader, from the chief executive to the newly employed warehouseperson.

There are certain words peculiar to each of our particular vocations which we have to use because no other words can adequately replace them. (Computer specialists, for example, have to refer to "disks," "disk drives," and "bits" and "bytes" of information.) To keep their sentences and paragraphs as uncluttered as possible, these technical words should be surrounded by mainly simple words.

Compare these two sentences:

A. An aberration of considerable magnitude significantly influenced the character readout.
B. A large deviation seriously affected the character readout.

Readers would need a very good vocabulary to understand all the words in the first sentence, and even then they would have to read carefully to fully grasp what is being said. Most readers would understand the second sentence.

Remove Words of Low Information Content

If you want your writing to create a positive, purposeful impact, you cannot afford to insert words and expressions which neither clarify nor contribute to the message. Such words are known as "low information

---

[1] Glen Leggett, and others: *Prentice-Hall Handbook for Writers*, 7th Edition, (Englewood Cliffs, NJ: Prentice-Hall, Inc., 1978).

content" (LIC) words, because they detract from rather than improve the message's clarity. There are several in this sentence:

> In order to effect an improvement in package handling an effort should be made to move the shipping department so that it is located in the vicinity of the loading dock.

The LIC words are:

> *In order to* (replace with *To*)
> *effect an improvement in* (use *improve*)
> *an effort should be made* (replace with *we should*)
> *located in the vicinity of* (use *near*)

Without the LIC words the sentence reads:

> To improve package handling we should move the shipping department so that it is near the loading dock.

Or (better still):

> To improve package handling we should move the shipping department nearer to the loading dock.

LIC words make one's writing seem woolly and indefinite. They flow easily from the tips of our pens and pencils, but then once they are on paper they can be hard to identify. For example:

When we have written	It can be difficult to think of
brings to a conclusion	concludes
for a period of	during
it will be necessary to	we must
in the direction of	toward

Simply being aware that you should not use LIC words in your reports will help you to be a careful writer, but still will not prevent you from inadvertently inserting them during an enthusiastic burst of writing. After you have written your first draft, but before the final copy is typed, always take a few minutes to check that you have not used any unnecessary words. The more common LIC words and expressions are listed in Table 10-1.

Eliminate Overworked Expressions

Overworked expressions can create a more noticeable negative effect than LIC words because they make their writer seem wordy or insincere, and sometimes pompous or evasive. Some typical expressions are listed in Table 10-2. These should be searched for, identified, and eliminated at the same time that you check for LIC words.

Table 10-1 Some Typical Low Information Content (LIC) Words and Expressions

These LIC words and phrases should be eliminated (indicated by X) or written in a shorter form (shown in parentheses).

actually (X)	in color; in length (long); in number; in size (X)
a majority of (most)	
a number of (many; several)	in connection with (about)
as a means of (for; to)	in fact; in point of fact (X)
as a result (so)	in order to (to)
as necessary (X)	in such a manner as to (to)
at present (X)	in terms of (in; for)
at the rate of (at)	in the course of (during)
at the same time as (while)	in the direction of (toward)
at this time (X)	in the event that (if)
bring to a conclusion (conclude)	in the form of (as)
by means of (by)	in the light of (X)
by the use of (by)	in the neighborhood of; in the vicinity of (about; approximately; near)
communicate with (talk to; telephone; write to)	
connected together (connected)	involves the use of (employs; uses)
contact (talk to; telephone; write to)	involve the necessity of (demand; require)
due to the fact that (because)	is a person who (X)
during the course of (during)	is designed to be (is)
during the time that (while)	it can be seen that (thus; so)
end result (result)	it is considered desirable (I or we want to)
exhibit a tendency to (tend to)	
for a period of (for)	it will be necessary to (I, you, or we must)
for the purpose of (for; to)	of considerable magnitude (large)
for the reason that; for this reason (because)	on account of (because)
	on the part of (X)
in all probability (probably)	previous to; prior to (before)
in an area where (where)	subsequent to (after)
in an effort to (to)	with the aid of (with)
in close proximity to (close to; near)	with the result that (so; therefore)

Table 10-2 Typical Overworked Expressions and Clichés

a matter of concern	in the foreseeable future
and/or	in the long run
all things being equal	in the matter of
as a last resort	it stands to reason
as a matter of fact	last but not least
as per	many and diverse
attached hereto	needless to say
at this point in time	on the right track
by no means	par for the course
conspicuous by its absence	please feel free to
easier said than done	pursuant to your request
enclosed herewith	regarding the matter of
for your information (as an introductory phrase)	slowly but surely
	this will acknowledge
if and when	we are pleased to advise
in reference to	we wish to state
in short supply	with reference to
	you are hereby advised

11
Writing a List of References or a Bibliography

Whenever you quote someone else's facts and figures, or draw information from a textbook, journal article, report, letter, or even a conversation, it is customary to acknowledge the source of your information within your report. This is usually done at the end of the report, in a section called References (or List of References), as Brian Lundeen has done in his computer analysis report in Chapter 8. References normally occur in longer reports and proposals—seldom in very short reports.

The purpose of a reference is threefold:

1. To give your report credibility (When readers encounter a statement such as "A previous study has shown that 26% of the city's core-area adults are unemployed," they expect to be told who made the original statement, and in what document it appeared.)
2. To help readers refer to the same source, if they want more information
3. To give credit to the originator

There are specific rules for writing a list of references, and to some extent they vary depending on where your report is to be published. The rules shown here are generally acceptable for any business or industry report. (But if, for example, your report is to appear in the journal of a professional society, then you should follow the style used by that journal.) I am assuming that most reports you write will be issued by your company or organization, and that a standard style will be most applicable.

HOW TO WRITE REFERENCES

References are listed in the sequence in which each item is mentioned in the report. If the first statement that needs to be supported concerns the quantity of water consumed by your city, then the first item in your list of references will be the document in which water consumption is tabulated. Each reference entry is numbered sequentially, starting at "1," and a corresponding number is shown in the report narrative to direct the reader's attention to the appropriate entry in the list of references. Here is an example:

> Over the past eight years the city's water consumption has ranged from a low of 207,389 gallons per day to a high of 253,461 gallons.[1] *(This "1" refers to the first entry in the list of references.)*

Since every entry is numbered, a corresponding number must appear in the report narrative for each reference. (See Brian Lundeen's report, Chapter 8.)

Each entry in the list of references must supply sufficient information for the reader to clearly identify the document and be able to refer to it or order it. The primary information a reader needs is

- Who made the statement
- In what document it appeared, or where and to whom it was said (if a spoken reference)
- When the statement was made

It's essential that details such as the author's name and document title be copied exactly as they appear on the original document, so that readers will experience no difficulty in finding or ordering it.

The preferred methods for listing the more common documents and conversations are described below.

Book by One Author. The entry should contain

> Author's name
> Book title (underlined)
> City of publication
> Name of publisher   (enclosed within parentheses)
> Date of publication
> Page number of specific reference (if applicable)

Using this book and page as an example:

1. Ron S. Blicq, <u>Guidelines for Report Writers</u> (Englewood Cliffs, NJ: Spectrum Books, 1982), <u>p. 152</u>.

Book by Two Authors. Both authors are named; all other information is the same as for a single-author book:

2. Bayard O. Wheeler and Thomas J. Adams, <u>The Business of Business: An Introduction</u> (San Francisco, CA: Canfield Press, 1973), p. 177.

Book by Three or More Authors. Only the primary author is named (usually the first-named author); remaining authors are replaced by the expression "and others." All other information is the same as for a single-author book:

> 3. Charles T. Brusaw and others, <u>The Business Writer's Handbook</u> (New York, NY: St. Martin's Press, 1976), p. 401.

Book Containing Sections Each Written by a Different Author, with the Whole Book Edited by Another Person. (Often applicable to anthologies.) If your reference is to the whole book, the editor's name is used and that person's editorial role is identified by the word "ed." immediately after his or her name:

> 4. Waris Shere, ed., <u>In Search of Peace</u> (Hauppage, NY: Exposition Press, 1980), p. 165.

If your reference is only to an article or section in the book, the author's name and section title are used:

> Author's name (or authors' names)
> Section title (in quotation marks)
> Book title (underlined)
> Editor's name (if the book has an editor)
> City of publication
> Name of publisher   (enclosed within parentheses)
> Date of publication
> Page number article begins, or of specific reference

For example:

> 5. Alexander M. Haig, Jr., "Nato's Strategy: The Challenge Ahead," in <u>In Search of Peace</u>, ed. Waris Shere (Hauppage, NY: Exposition Press, 1980), p. 23.

Book: Second or Third Edition. If a book is a second or subsequent edition, the words "2nd ed." (or 3rd ed., etc.) should be entered immediately after the book title.

Article in a Magazine or Journal. The entry should contain

> Author's name (or authors' names)
> Title of article (in quotation marks)
> Title of magazine or journal (underlined)
> Volume and issue numbers (shown as numerals only, e.g. 17:4)
> Magazine or journal date
> Page number on which article starts, or of specific reference

For example:

> Gerald L. Ratliff, "Performance Guide for Oral Communication" in <u>IEEE Transactions on Professional Communication,</u> PC-23:1, March 1980, p. 11.

If the author of a magazine article is not identified, the reference should start with the article title.

Report Written by Yourself or Another Person. The entry should include

> Author's name, or authors' names (if authors are identified)
> Title of report (underlined)
> Report number or identification (if applicable)
> Name and location of organization issuing report
> Date of report
> Specific page number (if applicable)

The report in Chapter 8 would be listed like this:

> 7. Brian E. Lundeen, <u>Evaluation of Computer Systems for The Western Farm Implement Company.</u> Report No. E-28, Antioch Business Consultants, Foothills, CO, December 19, 1981.

Technical Paper Presented at a Conference. The entry should contain

> Author's name (or authors' names)
> Title of paper
> Name of conference and sponsoring organization
> Location of conference
> Date of presentation

For example:

> 8. Robert C. Cornwell, <u>Human Needs and Their Impact on Communication,</u> Association of Records Managers and Administrators Conference: Information Management for the 80's, Boston, MA, October 21, 1980.

Letter or Correspondence. The entry should have

> Author's name
> Author's identification (employer and location)
> Form of correspondence (letter, memorandum, telegram)
> Addressee's identification (employer and location)
> Date of letter

For example:

> 9. Jacalyn Ainslie, Spartacus Equipment Inc., San Diego, CA. Letter to Winston Feldman, Nor'West Insurance Company, Denver, CO, January 13, 1982.

Speech or Conversation. The entry should be

> Speaker's name
> Speaker's identification (employer and location)
> Form of communication (speech, conversation, telephone call)
> Listener's name (individual or group)
> Listener's identification (employer and/or location)
> Date of communication

Writing a List of References or a Bibliography

Examples are:

10. Della A. Whittaker, Harry Diamond Laboratories, Washington, DC, speaking to the 28th International Technical Communications Conference, Pittsburgh, PA, May 21, 1981.
11. Gavin R. Johnstone, Lakeshore Industries, Minneapolis, MN, in conversation with Anna King, H. L. Winman and Associates, Cleveland, OH, December 13, 1981.

Second Reference to a Document. When a document is referred to more than once, an abbreviated reference containing only the author's surname (or authors' surnames) and new page number can be used for all subsequent entries. If, for example, further references are made to the documents listed earlier as entries 2 and 6, the new entries would be:

12. Wheeler and Adams, p. 104.
13. Ratliff, p. 63.

(Note that the Latin terms *ibid.* and *op. cit.* are not used in modern reports.) If several documents by the same author are referenced, then the date of publication is included in subsequent entries to identify which of the author's specific works is being referred to:

14. Carter, 1981, p. 147.

BIBLIOGRAPHIES AND FOOTNOTES

A bibliography is used when a report writer wants to list more documents than are referred to in the report. It becomes a comprehensive list of all documents pertaining to the topic being discussed, or which have been used to research and conduct the project or study.

In a bibliography the documents are listed in alphabetical order of authors' surnames, are not numbered, and the first line of each entry protrudes about five typewriter spaces (roughly half an inch) to the left of the remaining lines. The documents listed earlier as references are shown rearranged into a bibliography in Figure 11-1. Like a list of references, a bibliography appears at the end of the report narrative, but before the attachments or appendices.

Footnotes, which were once popular, now are not recommended for business and technical reports. Their position at the foot of the page on which each reference is made not only distracts readers' eyes and interrupts reading continuity, but also creates difficulties for the person typing the report. Footnotes are better replaced by *endnotes* (a list of references at the *end* of a report) which much more conveniently and unobtrusively serve the same purpose.

BIBLIOGRAPHY

Ainslie, Jacalyn, Spartacus Equipment Inc., San Diego, CA. Letter to Winston Feldman, Nor'West Insurance Company, Denver, CO, January 13, 1982.

Blicq, Ron S., *Guidelines for Report Writers* (Englewood Cliffs, NJ: Spectrum Books, 1982).

Brusaw, Charles T., and others, *The Business Writer's Handbook* (New York, NY: St. Martin's Press, 1976).

Cornwell, Robert C., *Human Needs and Their Impact on Communication*, Association of Records Managers and Administrators Conference: Information Management for the 80's, Boston, MA, October 21, 1980.

Haig, Alexander M. Jr, "Nato's Strategy: The Challenge Ahead," *In Search of Peace*, ed. Waris Shere (Hauppage, NY: Exposition Press, 1980).

Lundeen, Brian E., *Evaluation of Computer Systems for The Western Farm Implement Company*. Report No. E-28, Antioch Business Consultants, Foothills, CO, December 19, 1981.

Ratliff, Gerald L., "Performance Guide for Oral Communication," *IEEE Transactions on Professional Communication*, PC-23:1, March 1980.

Shere, Waris, ed., *In Search of Peace* (Hauppage, NY: Exposition Press, 1980).

Wheeler, Bayard O., and Thomas J Adams, *The Business of Business: An Introduction* (San Francisco, CA, Canfield Press, 1973).

Fig. 11-1. A typical bibliography formed from the references listed on these pages.

12
Inserting Illustrations into Reports

An illustration can help readers more readily understand a difficult part of a report, or a particular point a report writer wants to make. Because its role is to enhance rather than duplicate the narrative, an illustration must be simple, clear, and useful. A reader should not have to turn to the report's words to understand what an illustration depicts.

Illustrations appear mostly in longer, more formal reports, rather than in short informal reports. Hence, the illustrations described here are likely to be used in analyses, feasibility studies, proposals, and investigation or evaluation reports.

Before inserting an illustration a report writer should consider

1. Which kind of illustration (e.g. a table, graph, bar chart, flow diagram, photograph, etc.) will best illustrate a particular feature or characteristic.
2. Whether readers will be using the illustration simply to gain a visual impression of an aspect being discussed, or will be expected to extract information from it.
3. Whether the illustration will be referred to only once, to amplify or explain a point, or several times; if it will be referred to frequently, its position becomes important.

Every illustration should be numbered sequentially, so that it can be easily referred to in the report narrative, like this:

> ... in Figure 2 the decreasing monthly profits experienced in financial year 1981–82 are compared with profits for similar recessionary periods.

The illustration should also have a title, such as:

> Fig. 2. Financial year 1981–82 profits compared to two previous recessionary periods.

Sometimes a caption follows the title, to draw attention to an important point or explain some aspect in more detail. For example (continuing from the Figure 2 title, above):

> Both curves for previous periods show a distinctive flattening about six months before the profits reach their lowest level. This flattening is not evident in the 1981–82 curve.

Report writers must also decide whether an illustration should be placed directly in or beside the report narrative, or as an attachment or appendix at the end of the report. Here are some guidelines:

- If the illustration is extremely complex or fills more than one page, it should be an attachment.
- If readers will need to refer to the illustration as they read the report, it should be placed in the report narrative.
- If an illustration meets both the previous criteria, then the complete illustration (such as a large, extremely complex flow chart) should be an attachment, and a smaller, simpler, much less detailed illustration should be prepared and inserted into the report proper.

TABLES

Tables document information in tabular form, such as results of tests, quantities of items manufactured, daily receipts, etc. Unlike many of the illustrations described in this chapter, tables are meant to be examined in detail by the reader, who may want to extract or extrapolate data from them. Consequently the rules for preparing tables differ from the rules for preparing illustrations such as graphs and charts.

Guidelines for Preparing Tables
1. Keep the table simple, using as few columns as possible.
2. Limit the amount of data by omitting any details readers will not need.
3. Insert a clear, simple, but fully understandable title at the head of each column.
4. Insert a unit of measurement at the head of a column, rather than repeat the unit after each entry within the column (see how this has been done for "mph" and "$" in the table in Figure 12-1).
5. Insert the table number and an informative title, and center them either immediately above or below the table (above the table is preferred; see Figure 12-1).
6. Decide whether the table is to be "open" (without ruled lines separating the columns, as in Figure 12-1), or "closed" (with the ruled lines inserted).

The narrative part of the report should inform readers what they should learn from the table, so that its relevance is clear.

Table 2
EXCESS-SPEED DRIVING INFRACTIONS BY EMPLOYEES
OPERATING COMPANY VEHICLES
JANUARY 1 TO MARCH 31, 19xx

Date of Offence	Driver	Vehicle (Lic No.)	Speed Recorded (mph)	Speed in Excess of Limit (mph)	Fine ($)
Jan 17	B. Bastian*	JGR 611	101	36	41.00
Feb 04	K. Wall	JGR 608	66	26	26.50
Feb 04	M. Johnston	MND 147	61	11	18.50
Feb 19	R. Kurilicki	JGR 608	68	28	31.00
Mar 02	B. Bastian*	JGR 611	123	63	88.00
Mar 15	K. Fallis	MND 176	67	27	29.50
Mar 21	R. Ingraham	JGR 609	94	29	32.50

*Employment terminated March 31.

Fig. 12-1. An "open" table (no lines separate the columns of data)

GRAPHS

Graphs offer a simple way to illustrate how one factor affects, or is affected by, another. They have the particular advantage that, providing they are not too complex, the changes they depict can be readily visualized and understood by most readers. For example, graphs can be used to show

- Predicted sales for varying price structures
- The increase in fuel consumption resulting from increased highway speeds
- Personal life expectancy in relation to the quantity of cigarettes smoked
- Weight reduction in relation to lower calorie intakes

No matter how "technical" the subject being illustrated, graphs must be kept simple: their purpose is to quickly enhance the reader's understanding of the report. The guidelines listed below all contribute to this cardinal rule:

1. Limit the number of curves on a graph to three if the curves cross one another, or four if they do not intersect or there is only a simple intersection. If you have to construct a multiple-curve graph containing six or more curves, construct two graphs rather than one.
2. Differentiate between curves, particularly if they intersect, by assigning them different weights. Make the most important curve a bold line, the next most important a light line, the third curve a series of dashes, and the least important curve a series of dots (for an example, see Figure 12-2). Avoid the temptation to use colors to differentiate between curves, because all the colors will print black on any copies made with an office copier.
3. Position the curves so they are reasonably centered within the frame provided by the graph's axes. If necessary adjust the starting point of the scale(s) to move an off-centered curve to a more central position. See Figures 12-3 and 12-4.
4. Select a scale interval for each axis which will help the curves create a

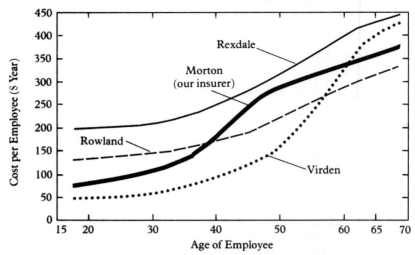

Fig. 12-2. A graph with four curves. The most important is identified by a bold line.

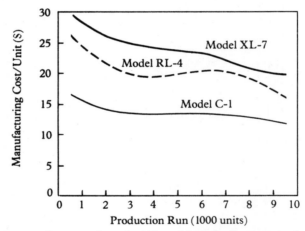

Fig. 12-3. An incorrectly centered graph. Although technically accurate, the graph appears unbalanced.

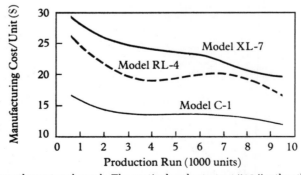

Fig. 12-4. A correctly centered graph. The vertical scale starts at "10," rather than at "0" as in figure 12-3.

visually accurate image (Figure 12-5). An incorrect scale interval may inadvertently cause curves to depict a false impression, as shown in Figure 12-6.

5. Omit all plot points, to provide a clean, uncluttered illustration. The only time that plot points or lines should be present is in a detailed drawing placed in an attachment, from which readers are expected to extract information or examine the graph's construction.

6. Keep all lettering clear, brief, and *horizontal*. The only nonhorizontal lettering should be along the vertical axis, as shown in Figure 12-2. Particularly avoid placing lettering along the slope of a curve.

7. Omit a grid unless you expect your readers will want to extract their own figures from the graph (compare the no-grid graph in Figure 12-4 with the gridded graph in Figure 12-5). If you are unsure whether to insert grid lines, you can insert an "implied" grid, as in Figure 12-2.

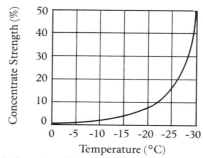

Fig. 12-5. Properly balanced scale intervals produce a visually accurate curve.

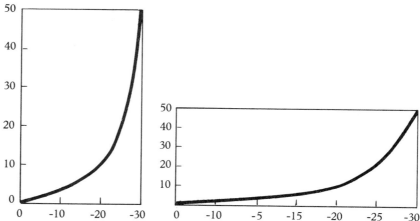

Fig. 12-6. The effect of an improperly balanced scale interval. Although these curves are technically accurate, neither creates the same visual impression as the correctly balanced curve in Figure 12-5. The contracted axes overaccentuate the "flattening" at one end, and deemphasize the "flattening" at the other end of each curve.

BAR CHARTS

Whereas graphs have two continuously variable functions, bar charts have only one. They are simpler to read and understand than graphs, and so are particularly useful as illustrations for nonspecialist, or lay, readers. Normally they provide only a general rather than a specific indication of results, quantity, time, etc., which makes them unsuitable for depicting exact units of measurement; and neither can readers extrapolate detailed or exact information from them.

A bar chart offers comparisons, using parallel bars of varying length to portray weight, growth, cost, life expectancy, etc., of various items. The bars are arranged either vertically or horizontally, depending on the factors being displayed, with the variable function lying along the axis which runs parallel to the bars.

General Guidelines for Preparing Bar Charts
1. Position the bars so they are spaced one bar width apart.
2. Arrange the bars vertically when you are portraying "growth" factors, such as weight, quantity, cost, or units produced (see Figure 12-7).
3. Arrange the bars horizontally when you are portraying elapsed time, or factors in which time is a significant element; i.e. life expectancy, production time, project length, as shown in Figure 12-8.

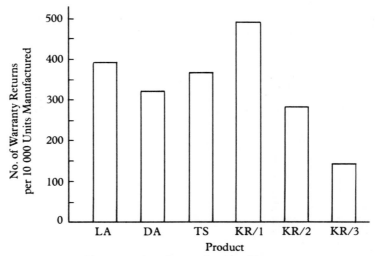

Fig. 12-7. A bar chart with vertical bars.

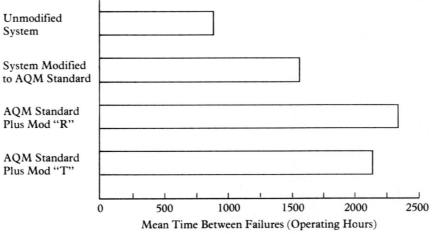

Fig. 12-8. A bar chart with horizontal bars.

4. Shade or cross-hatch the bars if you need to make them stand out.
5. If it is important for readers to know the exact total each bar represents, show the totals either immediately above the tops of the bars (if the figures are short enough) or inside the bars (along their length), as shown in Figure 12-9.
6. If the bars are composed of several segments, either identify the segments by various types of shading (and provide a legend beside or below the chart) or, if there is room, identify each segment with a word or two inside the bars (see Figure 12-10).

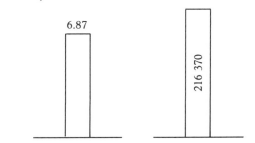

Fig. 12-9. Numbers placed above or within bars show exact figures.

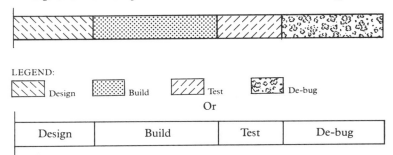

Fig. 12-10. Bars can be divided into segments either by shading or by lettering.

An unusual illustrative technique is to use pictorial "bars," inserting a picture of a car, person, tree, etc., the size or height of which provides the comparison (see Figure 12-11). This technique is seen more often in magazines and newspapers than in business reports.

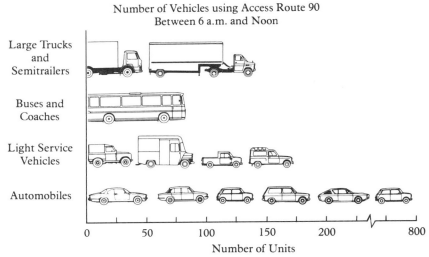

Fig. 12-11. A pictorial bar chart. The horizontal axis is broken and shortened between 225 and 775 to permit the very much longer automobiles "bar" to be depicted without unbalancing the illustration.

HISTOGRAMS

A histogram contains some features common to both a graph and a bar chart. It has two continuously variable functions, but is constructed like a bar chart because there are insufficient data on which to plot a true curve. To show that it has this dual but limited function, the bars are plotted immediately against each other as in Figure 12-12. Indeed, a line drawn through the tops of the bars would produce a rudimentary curve.

Guidelines for preparing a histogram are similar to those applicable to preparing a graph and a bar chart.

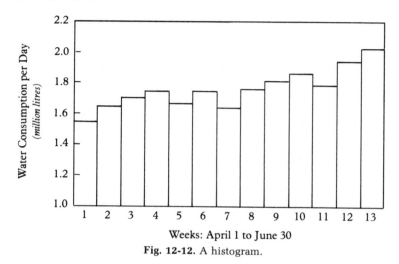

Fig. 12-12. A histogram.

SURFACE CHARTS

A surface chart seems to combine the characteristics of both a graph and a histogram. It has two continuous variables, and it is made up of adjoining bars from which the vertical construction lines have been erased (see Figure 12-13). But the curves, instead of being compared as in a graph, are summated. That is, for each vertical "bar," the factors being displayed are added, so that the first curve *becomes the base* for the second curve; and the second curve *becomes the base* for the third curve. Thus the uppermost curve represents the total of all the curves added together.

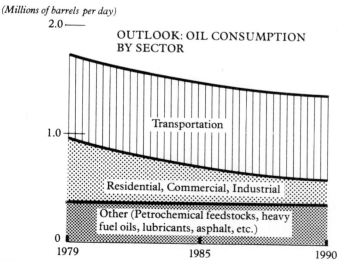

Fig. 12-13. A surface chart (courtesy of *The Financial Post*).

Readers cannot easily extrapolate information from surface charts because direct readings can be extracted only from the lowest and uppermost curves.

Guidelines for Preparing a Surface Chart
1. For the lowest curve, choose and draw in the factor which is the most important, represents the largest quantity to be depicted, or offers the most stable (even) curve.
2. For the second curve, select the next factor and plot it in, using the first curve as the base for each section.
3. Repeat the sequence for each additional curve.
4. Shade or cross-hatch the curves, preferably making the lowest section the darkest and the uppermost section the lightest.

PIE CHARTS

A pie chart is one of the simplest forms of illustration. By dividing a circle (a "pie") into segments of varying size, it can show market distributions, tax apportionment, product costs, in a readily understandable form. Because it is a simple illustration, only a few guidelines are necessary:

1. Always make the segments of a pie chart add up to an exact 1, 100% or $1 (or a round-figure multiple: $100, $1 million).
2. Check that the segments are visually accurate; i.e. are in the correct proportions for the quantities they depict.
3. If there are a lot of very small segments to depict, combine them into one segment and label it "Miscellaneous" (or use a more descriptive term). If it is important for readers to know the composition of this segment, provide a list beside the illustration or in a caption below the chart.

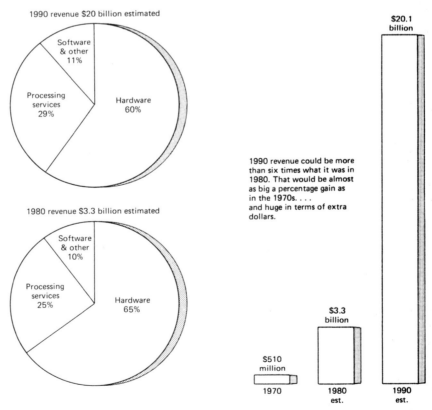

Fig. 12-14. A pie chart, combined with a bar chart to form a composite illustration (courtesy *The Financial Post*).

Figure 12-14 contains both a pie chart and a simple bar chart. Imaginative combination of charts in this manner can make a report much more visually appealing.

FLOW CHARTS, SITE PLANS, AND LINE DIAGRAMS

A flow chart (see Figure 12-15) provides a visual description of a procedure, process, plan or system. A site plan depicts the more significant features of a building site or small area of a town, while a line diagram can encompass anything which needs to be illustrated (e.g. a piece of equipment, hookup of several instruments, layout of an office, etc). In all cases they should

1. Be as simple as possible.
2. Clarify the accompanying written description
3. Contain only the essential elements (which means firmly eliminating unessential elements)
4. Be easy to follow
5. Be readily understood without having to read the written description
6. Be drawn in clear black ink
7. Contain neatly lettered, clear, but brief, explanatory words

Brian Lundeen's computer investigation report in Chapter 8 contains a flow chart.

PHOTOGRAPHS

Photographs are an ideal, accurate way to show readers either close-up details or "the whole picture," but unfortunately they are awkward and expensive to reproduce. Consequently report writers have to consider whether the time and cost of reproducing photographs is justified, or if the same information can be conveyed just as easily and more economically by inserting a sketch or line drawing into a report.

The chief problem with photographs is that the average office copier does not reproduce them well. To achieve a quality image, photographs have to be carefully prepared for printing by a lithographer, who places a fine dot screen over them and makes a special plate for printing them on bond paper with an offset duplicator. As few businesses have the facilities to do all of this in-house, the work usually has to be subcontracted. This can be inconvenient and costly unless a large number of copies of the report are to be printed.

If only a very few copies are to be made of a report, photographic prints can be inserted manually in several ways:

1. If there are several prints and they are small, they may be glued into the report in spaces left open for them during typing.
2. If there are only one or two prints and they are large (8 × 10 in.), they may be glued onto a full-size page on which a title and caption have been pretyped.
3. If there are several medium or large prints, they may be inserted into a prepared pocket or sleeve (an envelope with one end snipped off is ideal), which has been mounted on a piece of card. The pocket is placed at the back of the report and preceded by a sheet listing the photographs. Each photograph should be coded on the front for identification, and cross-referenced to this list.

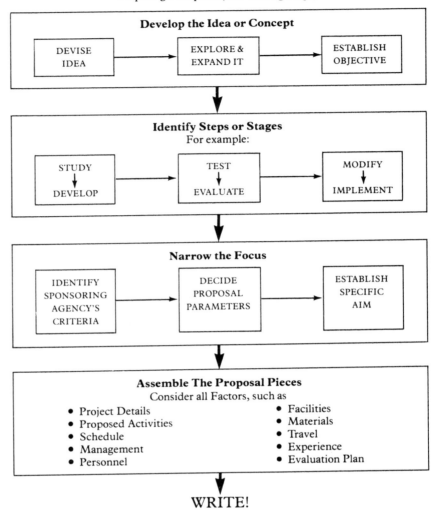

Fig. 12-15. A flow chart (also known as a flow diagram).

Remember that photographs are bulky and tend to curl. If there are too many of them they can create an uneven, awkward-looking document, in which case you may prefer to bind them into a separate folder which accompanies your report.

THE SIZE AND POSITION OF ILLUSTRATIONS

The ideal illustration sits beside or immediately above or below the words which refer to it. Unfortunately this convenient juxtaposition is not always easy to achieve, particularly if a report is to be printed on only one side of the paper. If full-page tables and illustrations are inserted into the narrative, they completely interrupt reading continuity.

Use these guidelines when preparing a report for typing:

1. Plan the report's pages before they are typed in their final form, even if doing so means having an intermediary draft typed so that you can evaluate how much space each paragraph will require.

2. Keep diagrams as simple as possible and as small as possible so there will be room to insert type above or below them.
3. Position diagrams so they are adjacent to the paragraphs which refer to them or to paragraphs which most need illustrative support.
4. Beneath every illustration insert the figure number; a brief title; and, possibly, an explanatory caption (see how this has been done in Figure 12-16).
5. If a full-page diagram has to be inserted, consider whether it must accompany the report narrative, or can be placed in an attachment with a small, simple sketch inserted in its place in the body of the report.
6. If a full-page diagram is horizontally oriented (i.e., its base is longer than its height), turn it through 90° so that it will be read from the right-hand side of the page. (See Figure 12-16, and the horizontally oriented table in Attachments 1 and 2 of Lorraine Dychuk's proposal in Chapter 7.) This orientation applies regardless of whether the illustration is in the body of the report or is an attachment or appendix.
7. Check that every illustration is referred to in the report narrative, by either its figure number or its attachment identification.

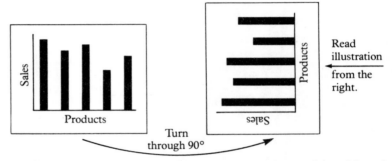

Fig. 12-16. Full-page horizontal diagrams are turned so that they can be read from the right. This should be done even though some words may be inverted when the illustration is viewed from the foot of the page.

13
Guidelines for Spelling, Abbreviations, and Numbers

Every publisher has a style manual which establishes rules for spelling, capitalization, writing numbers, and so on. The publishers' writers and editors refer to the style manual as their bible, and use it whenever a decision is necessary (e.g. when they need to know if they should write "twelve" or "12," "programed" or "programmed," "MPH" or "m.p.h." or "mph"). As individual report writers we do not have a special style manual to refer to, but we can adopt some of the basic guidelines which are common to most style manuals. These, and additional suggestions, are outlined here.

SPELLING

Select a U.S.-published dictionary which is large enough to carry most of the words you are likely to use (a 600-page dictionary is ideal), and which has been revised within the past five years (so that it reflects current spelling practice). Use this dictionary's spelling rules as *your* rules, so that you will be consistent.

Where the dictionary shows alternatives, such as "center, centre" and "programer, programmer," generally choose the first-listed spelling because dictionary makers normally list the preferred or more common spelling first. If you decide you prefer the alternative spelling, then underline it with a colored pen so you will be reminded of your preference when you look up the word later. The need to identify an alternative spelling may seem rare, but there will be occasions when you feel that the first-listed spelling is awkward (for example, many people feel "programer" to be an unnatural spelling, and choose to write "programmer").

Some useful books which list many "problem" words are: Rudolph Flesch's *Look it Up*[1]; The *Government Printing Office Style Manual*,[2] which lists nearly 20,000 compound terms; and, for technical report writers, *Technically-Write!*,[3] which lists many technical words in its glossary.

ABBREVIATIONS

You may abbreviate any term you wish, particularly if it is a lengthy term which will be used frequently in a report. But there are guidelines you should adhere to:

1. Indicate to the reader what your abbreviation means by stating it fully the first time and then showing the abbreviated form in parentheses beside it:

The digital reference number (drf) we applied to the first set . . .
. . . in the second set, the drf was determined by . . .

2. Avoid forming your own abbreviation when another abbreviation already exists and is commonly used.

3. Use lower case letters, unless the abbreviation is formed from a proper noun such as a person's name:

foot, feet	ft
average	ave
ampere	A (formed from Ampere)

4. Omit all punctuation, unless the abbreviated term forms another word:

inch	in.
absolute	abs
approximately	approx

5. Omit the *s* from an abbreviation of a plural quantity:

feet	ft
inches	in.
pounds	lb
hours	hr

Be aware that there are some exceptions to these guidelines, caused by nonstandard terms being adopted through general usage. For example, although "no." is the correct abbreviation for "number," you are much more likely to see it used as "No." (and sometimes even as "#," which is definitely nonstandard); hence, you will not be wrong if you choose to use "No." But be consistent: whichever term you use, use it all the time.

[1] Rudolph Flesch, *Look it Up: A Deskbook of American Spelling and Style* (New York: Harper and Row, 1977).
[2] *United States Government Printing Office Style Manual*, Superintendent of Documents, Washington, DC 20402.
[3] Ron S. Blicq, *Technically-Write! Communicating in a Technological Era*, 2nd ed. (Englewood Cliffs, NJ, Prentice-Hall, Inc., 1981), p. 341.

NUMBERS

Publishers' style manuals also reflect standard usage for writing numbers when they appear as part of a sentence or paragraph. The general rule is:

> From one to nine, spell the number.
> For 10 and higher, use numerals.

But there are exceptions to this rule, and sometimes you may have to decide whether the rule or the exception takes precedence. The exceptions are:

1. Spell out the number if it is
   - the first word in a sentence
   - a large generalization (as in "...about eight thousand...")
   - a fraction which is less than one (as in "...only one third of the participants...")
2. Use numerals if the number
   - is part of a series of quoted numbers
   - is a year, date, time, age (of a person), percentage, or sum of money
   - is part of a unit of measurement (as in "28 lb")
   - is specific technical data, such as a dimension, tolerance, temperature, or result of a test
   - contains a decimal or fraction (as in "6½" and "3.25")
   - refers to a chapter, figure (illustration), or page (as in "page 128")

It is also standard practice to insert a zero at the start of any decimal which is less than "1" (as in "0.25" and "0.0056").

METRIC (SI) UNITS

If metrication becomes more firmly established in the United States, as it is worldwide, report writers—and particularly those who write technical reports—will have to know the rules for writing metric (SI) units. These were defined originally by the eleventh (1960) Conférence Générale des Poids et Mesures (CGPM) in Paris, France, and have been only slightly revised in the intervening years. Very briefly, the current guidelines for writing SI symbols are:

1. Use upright (*not italic*) type.
2. Use lower case letters, except when a symbol is derived from a person's name (as in "V" for "volts," which is derived from "Voltaire").
3. Insert a space between the numeral and the first letter of the symbol (as in "38 kHz"), but *no* space between the symbols themselves (e.g. there is no space between "k" and "Hz").
4. Omit the 's' from all plurals (as in "221 km"), and do not insert a period at the end of the symbol (except when the symbol is the last word in a sentence).
5. Insert an oblique stroke for the word "per" (as in "km/hr"), and a dot at mid-letter height to show that two symbols are multiplied (as in "N·m," for "Newton meter").

More detailed instructions, plus numerous examples, are contained in the glossary (Chapter 11) of *Technically-Write!*, referenced earlier.

14
Guidelines for Working with a Report Production Team

A report writer's work is not over when he or she hands in a handwritten draft for typing. From that moment on, the report writer becomes part of a report production team, with each member being responsible for certain aspects of the report. The writer has overriding responsibility not only for the information conveyed in the report, but also for the quality of language and the report's appearance.

A report production team can be composed of as few as two persons or as many as seven or eight, or even more. In a small company the production team may comprise only the report writer and the typist, with one of them making copies of the report on an office copier. In a large company the production team may include one or two illustrators, two or more typists, a printer, and an editor or publications supervisor, all in addition to the report writer. (When the team has an editor, responsibility for correctness of language and the report's appearance usually shifts to that person. The report writer retains responsibility for correctness of information, and should share some responsibility for correctness of language.) If more than one person writes a report, and there is no editor, one of the writers is normally appointed coordinating writer.

The image that readers perceive, both of the report writer and of the company or organization he or she works for, is directly influenced by what they see and read. Words poorly centered on the title page, typing errors, misspelled words, unevenly positioned page numbers, and grammatically incorrect sentences create an image of a sloppy worker employed by an organization which produces a low-quality product or service. On the other hand, good language and crisp, clear typing neatly positioned on every page convey the image of a confident report writer employed by a high-quality organization.

Being responsible for the production of your own report means working cooperatively with the remaining team members, principally the persons who will type, illustrate, and print it. The guidelines listed below suggest ways for achieving a harmonious atmosphere.

Your first step should be to set up a schedule, and then to meet with all the persons who will be involved in the report to check that the dates you have allowed for each stage are realistic for the amount of work required. The schedule may be a simple list showing proposed dates of completion for each stage, or it can be a form that travels with the report, as illustrated in Figure 14-1.

REPORT PRODUCTION CONTROL SHEET					
Step	Action	Planned Compl Date	Action by	Init.	Date Step Completed
1	Write first draft	Feb 16	Author	DML	Feb 15
2	Type first draft	Feb 26	Typist		
3	Check typed draft	Feb 28	Author		
4	Prepare illustrations	Feb 28	Drafting		
5	Check illustrations	Mar 1	Author		
6	Check draft & illus.	Mar 4	Supervisor		
7	Make final revisions	Mar 6	Author		
8	Type second draft*		Typist		
9	Check second draft*		Author		
10	Type final report	Mar 11	Typist		
11	Check final report	Mar 12	Author		
12	Print report copies	Mar 15	Printing		

* Steps 8 & 9 apply only if many changes are made to the first draft.

Fig. 14-1. A report schedule, to accompany a report.

WORKING WITH TYPISTS

Always give typists clear guidelines. Unless you tell them exactly what you have in mind they will type your words in a format they are accustomed to, which may not coincide with your ideas. Typists need to know:

1. The size of the typing area or, alternatively, the width of margins (e.g. 1⅛ in. minimum each side and at the top of the page, with 1½ in. at the bottom to allow for the page number).
2. Whether the report is to have single, 1½, or double line spacing.
3. The number of blank lines between paragraphs.
4. Whether the first line of each paragraph is to be indented, or the paragraphs are to be typed solid against the left-hand margin. (Marjorie Franckel's progress report on page 39 is the only report which has been typed with the first line of each paragraph indented; all other reports have been typed "full block.")
5. Whether the typed original or a printed copy will be sent to the addressee. (If the original is to be sent, then the typist cannot use much white corrective fluid to erase errors; if printed copies are to be distributed, and the original is to remain with the report writer, then correcting fluid can be used.)
6. The pitch and type style you prefer. You have a choice of type size if the typewriter has dual pitch (i.e. 10 [pica] or 12 [elite] characters to the inch).
7. Whether the version being typed is a draft or the final.
8. Where spaces are to be left for illustrations, and the exact amount of space required.
9. The date the typing is required.

You also need to work out who is to proofread the typed work. Proofreading should be done by two people, with one reading aloud to the other. This can be two typists, or you and a typist. Proofreading alone is both tedious and prone to error: it's much too easy not to notice a missing line or phrase.

WORKING WITH A WORD PROCESSING SYSTEM

Working with a word processing (WP) system is similar to working with typists, but much faster. Its main advantage is that once you have had the first draft of a report typed, you can make changes to it and have a second draft, with all the corrections incorporated, in your hands within minutes of giving the edited first draft to the typist.

A word processor consists of a modified typewriter keyboard, normally with a calculator keyboard beside it and a video screen above it. The typist types your draft onto the keyboard, as is done with a regular electric typewriter, and each line of type appears on the video screen until some 16 to 20 lines are visible. The typist then checks the words on the screen and corrects any errors by moving a small cursor over each letter to be corrected, and typing in a new letter. If several words have been missed, they are typed in at the correct place. The WP system then automatically shuffles the existing words along and onto the successive lines, to maintain the prescribed column width. (The WP system can also automatically justify the right-hand margin so that every line of type is exactly the same length.)

When the whole report has been typed, and then checked and corrected on the video screen, the typist pushes a *print* button on the keyboard. The report is then rapidly printed onto paper at some 60 to 120 characters per second. The machine that does this is known as a line printer, and it may stand on a table beside the typist or be kept in a separate room. Meanwhile the computer section of the WP system stores the report in its memory.

When you have made your corrections on the printed sheets, the typist recalls the report from the computer's memory (this takes only one to three seconds when a "floppy disk" is used to store your words), displays the report on the video screen, and types in the corrections. A few minutes later you have the corrected report to work with.

The speed with which you can see a corrected copy or the final report is the main advantage of a WP system. A secondary advantage accrues if you can type reasonably accurately yourself, for then you can type your first draft directly into the computer rather than laboriously handwrite it.

Most of the guidelines for working with typists who are using a standard electric typewriter are equally valid for typists using a WP system. The exceptions are:

1. The typist needs to know if the right-hand margin is to be justified, and how wide the lines of type are to be.
2. There is no need to use correcting fluid (previous Guideline No. 5).
3. There is less likely to be a choice of typewriter pitch (previous Guideline No. 6).

WORKING WITH ILLUSTRATORS AND DRAFTSPERSONS

Top-notch communication is extremely important between a report writer and an illustrator or draftsperson, who for clarity I shall refer to jointly as "illustrator." If the illustrations in your report are to complement the words you have written, your illustrator needs to know: something about the topic; the purpose of the report; who the reader(s) will be; and what aspects need to be emphasized.

Guidelines for effective communication between yourself and the illustrator are:

1. Explain the purpose of the report, and of each illustration.
2. Discuss how each illustration is to support your words (give the illustrator a draft copy of your report to read), and describe what parts are most important. If possible, sketch each illustration as best you can, and then give the sketches to the illustrator so he or she can visualize what you have in mind.
3. Provide accurate vertical and horizontal dimensions for each illustration. If the illustration is to be preceded or followed by typing, allow sufficient white space so that it does not look crowded.
4. Be sure to allow the illustrator plenty of time. Quote an illustration completion date (preferably in writing), and obtain the illustrator's assurance that the date can be met. Never say you want your illustrations ASAP (as soon as possible).
5. Discuss oversize drawings and by how much they will be reduced photographically, so that the illustrator will know not to make construction lines too light.

WORKING WITH A PRINTER

Most business and technical reports are printed in-house, or by a local "quick copy" service. In both cases the reports are duplicated on an office copier if only a few copies are required, or on an offset press if more than, say, 30 copies are required. (Only rarely is a report taken to a professional printer. Corporate annual reports would be a typical example.) If your report is printed in-house and the production run is short, you or your typist is likely to make the copies on a centrally located office copier. But if your organization has its own print shop, or if you use a "quick copy" service, then you should discuss the job with the person who does the printing.

Guidelines for Working with a Printer

1. See the printer before the final typing and illustrating are done, and discuss the report. Find out what equipment the printer has, and if there are any special requirements or limitations (e.g. whether the printer can reduce a large drawing photographically and print paragraphs or drawings which have been glued onto the typed original without creating shadow lines). Mention the date you plan to bring the job in, and ask how long the printing will take.

2. If the report is large or many copies are required, visit several printers and ask for cost estimates. Be sure to give the same requirements to each printer (e.g., number of pages and number of copies, how many photographs are being used, and whether the printer is to collate and bind the report).

3. When you take the job in for printing, write clear, complete instructions to the printer and clip or pin them to the job. Your instructions should include
   - Number of copies required
   - Color of ink to be used
   - Weight of paper (e.g. 20 lb bond)
   - Size of paper (e.g. 8½ × 11 in.)
   - Whether the report is to be printed on only one side or both sides of the paper
   - Where photographs and drawings are to be inserted (they usually travel separately, in an envelope)
   - Whether the whole job is to be reduced photographically, and by how much (stated as a percentage: "Reduce by 10%" or "Reduce to 85%")
   - Whether the job is to be collated and bound, and the type of binding required
   - Any special instructions
   - The date you require the printed report

4. If the report is at all complex, make a mock-up showing how the finished product should appear. Use the correct number of blank sheets of paper, fold them once, and staple them together to form a booklet. Open up the booklet and write a descriptive word on each page to show what should be printed there. If a page is to be left blank, write "BLANK PAGE" on it. The printer will use the booklet to determine what to print on each page, and to assemble the sheets in the correct sequence.

WORKING WITH MANAGEMENT

Unless your company employs a full-time technical editor, your draft reports will travel upward through the chain of command, with someone at each level adding, deleting, or changing words according to his or her whim. Unless you are very lucky, the report you receive back from management may not look like the document you wrote.

You can avoid some of this frustration by going to your manager or immediate supervisor and asking for guidelines. Tell your manager that you need to know:

1. For each report (a) who the primary reader is; (b) who the secondary readers are likely to be; and (c) who is most likely to use the information you supply, or take action as a result of your report
2. If you can use the "pyramid technique" (main message up front) for *all* of your reports, regardless of whether they contain good or bad news
3. If you can use the first person ("I" or "we") in your reports
4. If you can use the more emphatic active voice, rather than stick to the dull, less interesting passive voice

Finally, when you submit your draft report for evaluation, try to send a copy which is absolutely clean (i.e., has no penciled or ink alterations, or ragged typing corrections). A rough-looking report invites the reviewer to make more corrections, whereas a clean copy tends to inhibit the evaluator from marking up the report. As a second level of assurance, try attaching a sheet of pink paper to the front of the report, on which you have printed a request for report reviewers to make their suggestions *on the pink sheet*, rather than mark up the clean original. When reviewers have to spend time writing and cross-referencing every suggestion, they tend to write only the important ones.

Index

Abbreviations, forming, 170
Active voice, 145–47
Analysis, comparative (*see* Comparative analysis)
Appendices, 99, 100–101, 111–12, 129–130 (*see also* Attachments)
Attachments, as part of report:
   in formal report, 99, 101–101, 111–12, 129–130
   identified as Back-up compartment, (fig.) 36
   in investigation report, 45, 46, 64, 68, 73
   in laboratory report, 52, 54, 59, 60, 61
   in progress report, 36, 38, 41, 42
   in proposal, 78, 79, 80, 81, 84, 85, 92–93

Background, as part of report (*see also* Introduction):
   basic application, 13–14
   in informal suggestion, 78, 79
   in inspection reports, 27, 28, 30
   in short investigation report, 45, 46
   in laboratory report, 52, 60
   in occurrence report, 18
   in progress reports, 36, 37, 38, 40, 42
   in semiformal proposal, 80, 81, 82, 86–87
   in trip reports, 20, 21, 24
Back-up, as part of report (*see* Attachments)
Bar charts, 162–63
Bibliography, how to write, 155–6 (*see also* References)

Clichés, removing, 148, 149
Comparative analysis:
   in formal report, 11, 126–7
   in proposal, 80, 81, 84, 90–91
Compartments, for organizing writing (*see* Writing compartments)
Conclusions, as part of report (*see also* Outcome):
   in formal report, 99, 100, 101, 110–11, 127
   in investigation report, 64, 68, 72
   in laboratory report, 52, 54, 60, 61
   in semiformal proposal, 84, 91
Conference, reporting attendance, 26 (*see also* Trip Report)
Cover letter, 101–2, 113

Details, in basic report, 11–14
   development of, 11–13
   expansion of, 13–14
Diagrams (*see* Illustrations)
Dictionary, selecting, 169–70
Directness in writing, 141–47
   active voice, 145–47
   pyramid approach, 3–11, 142–43
Discussion, as part of report (*see also* Facts and Events):
   in formal proposal, 94, 95–96
   in formal report, 99, 100, 101, 104–10, 117–27
   in informal proposal, 80, 81, 82–84
Draftspersons, working with, 176

180 Index

Drawings in reports, 157–68 (see also Illustrations)

Editor, working with, 177
Evaluation report, 114 (see also Formal report)
Evidence, as part of report (see Attachments)

Facts and Events, as part of report (see also Discussion):
 basic application, 13–14
 in informal suggestion, labeled Details, 78, 79
 in inspection report, labeled The Findings, 26, 27, 28, 29
 in investigation report, labeled Investigation, 45, 46, 64, 65–68, 69–72
 in laboratory report, 51, 52–54, 60, 61
 in occurrence report, 18–19
 in progress report, labeled Progress, 36, 37, 38, 40, 42
 in trip report, labeled The Job, 20, 21, 22, 24
Feasibility study, 99 (see also Formal report)
Flow charts, 166, 167
Footnotes, 155 (see also References)
Formal report, 99–130
 appendix, 99, 100, 101, 111–12, 130
 arrangement of parts,
  alternative arrangement, 101
  traditional arrangement, 99–100
 circumstances for writing, 99
 comments on example, 101–112
 conclusions, 99, 100, 101, 110–11, 127
 cover letter, 101–102, 113
 discussion, 99, 100, 101, 104–11, 117–27
 example of report, 113–30
 introduction, 99, 100, 101, 104, 117
 recommendations, 99, 100, 101, 111, 128
 references, 100, 101, 111, 128
 report-within-a-report, 107, 108
 summary, 99, 100, 101, 103, 115
 table of contents, 100, 101, 103–4, 116
 title page, 100, 101, 102, 114
 writing compartments:
  alternative arrangement, 101
  traditional arrangement, 100

Graphs, 159–61
Guidelines for using this book, 3–4

Headings:
 in investigation report, 65
 in progress report, 41
Histograms, 164

Illustrations in reports, 157–68
 bar charts, 162–63
 example, in formal report, 106, 120
 flow charts, 166, 167
 graphs, 159–61
 guidelines for using, 157–58
 histograms, 164
 line diagrams, 166
 pie chart, 165–66
 photographs, 166–67
 positioning illustrations, 167–68
 site plans, 166
 size of illustration, 167–68
 surface chart, 164–65
 tables, 158–59
Illustrator, working with, 176
Informative writing, 141
Inspection report, 26–33
 arrangement of subcompartments,
  conditions found, 26, 27, 28, 30
  deficiencies, 26, 28, 29, 30, 32
 circumstances for writing, 26
 comments on examples, 28, 30, 32
 examples, 29, 31–32
 form for recording inspection data, 33
 similarity to trip report, 26
 writing compartments, 26–28
Introduction, as part of report (see also Background):
 in formal proposal, 94, 95
 in formal report, 99, 100, 101, 104, 117
 in investigation report, 64, 65, 69
Investigation report:
 formal report, 99–130
 letter report:
  comparison with semiformal report, 74–75
 semiformal report, 63–75
  circumstances for writing, 63
  comments on example, 64–68
  comparison with letter report, 74–75
  example, 69–73
  writing compartments, 63, 64, 65–68
 short report, 45–47
  circumstances for writing, 45
  comments on example, 46
  example, 47
  writing compartments, 45, 46

Laboratory report, 51–61
 academic lab report, 60–61
  circumstances for writing, 60
  organization plan, 60, 61
  writing compartments, 60–61
 industrial lab report, 51–60
  circumstances for writing, 51
  comments on example, 52–54
  example, 55–59
  prepared on a form, 52
  writing compartments, 52–54
Language of reports, 139
Letter report, 133, 134, 136–37 (see also Reports, appearance)
 format, 134, 136–37
 guidelines for typing, (fig.) 137, 137
Line diagrams, 166

Management editing, 177–78
Memorandum report, 133, 135 (see also Reports, appearance)
 format, 135
 guidelines for typing, (fig.) 135
Main message:
 focusing on, 9–11
 identification of, 7
 as summary statement, 10–12

Metric units, guidelines for writing, 171

Numbers in report narrative, 171

Occurrence report, 17–19
  comments on example, 18–19
  definition of, 17
  examples, 12–13, 14, 19
  writing compartments, 18
Organization plan:
  formal proposal, 94–96
  formal report, part of, 100, 108
  laboratory report, 61
  semiformal proposal, 82, 85
Outcome, as part of report (see also Conclusions and Recommendations):
  basic application, 13–14
  in informal suggestion, 78, 79, 80
  in inspection report, 26, 27, 30, 31
  in short investigation report, 45, 46
  in laboratory report, 52, 54, 67, 68
  in occurrence report, 18, 19
  in progress report, relabeled Plans, 36, 37, 38, 40, 42
  in trip reports, 20, 22, 24

Paragraph numbers, in progress reports, 41
Passive voice, 145–46
Pie charts, 165–66
Photographs, 166–67
Plan:
  for reading this book, 3–4
  for writing semiformal proposal, 82, 84–85
Printer, working with, 177
Progress report, 35–45
  comments on examples:
    occasional report, 38
    periodic report, 42
  description of, 35–36
  examples:
    occasional report, 39
    periodic report, 43–44
  headings, use of, 41
  paragraph numbering, 41
  occasional progress report, 37–39
  periodic progress report, 40–45
  writing compartments:
    occasional report, 37–38
    periodic report, 40–41
Project report, 99 (see also Formal report)
Proposal:
  formal proposal, 94–96
    circumstances for writing, 78, 94
    writing compartments, 94–96
  informal suggestion, 77, 78–80
    circumstances for writing, 77
    comments on example, 79
    example, 79–80
    writing compartments, 78
  semiformal proposal, 80–93
    circumstances for writing, 77
    comments on example, 81–85
    example, 86–93
    organization plan, 80–81, 84–85
    writing compartments, 80–81
    writing plan, 84–85

Punctuation in correspondence, 134
Pyramid method of writing, 5–14
  basic framework, 8
  details section:
    basic writing compartments, 13–14
    development of, 11–13
    example, (fig.) 14
    expansion of, 13–16
  example of report, (fig.) 14
  journalism, application to, 9–10
  organization of information, 8
  reports, application to, 10–11
  writing compartments, (fig.) 13
Pyramid, report writer's (fig.):
  basic pyramid, 13
  for formal report, 100, 101
  for inspection report, 26
  for semiformal investigation report, 64
  for short investigation report, 45
  for laboratory report, 52
  for occurrence report, 18
  for progress report, 36
  for formal proposal, 94
  for semiformal proposal, 80
  for reports with attachments, 36
  for informal suggestion, 78
  for trip report, 20

Reader, identification of, 4–5, 139–40
Reading plan, 3–4
Recommendations, as part of report (see also Outcome):
  in formal report, 99, 100, 101, 111, 128
  in investigation report, 64, 68, 72
  in proposal, 84, 91
References, writing a list of, 151–56
  bibliography, compared to, 155
  example, 111, 128, 129
  footnotes, compared to, 155
  guidelines for writing, 152
  purpose of references, 151–52
Reports:
  appearance, 133–8
    as a bound document, 134
    as a letter, 133, 134, 136–37
    as a memorandum, 133, 135
    as a titled page, 133, 136
    importance of good presentation, 133, 173–74
  examples of reports:
    formal report, 113–30
    inspection report, 29, 31–32
    investigation report,
      formal report, 113–30
      semiformal report, 69–73
      short report, 47
    laboratory report, 55–9
    occurrence report, 12–13, 14, 19
    progress report, 39, 43–44
    proposal, 86–93
    trip report, 23, 25
  format, 133–38
  illustrations and diagrams, 157–68 (see also Illustrations)
  language used in reports, 139
  main sections of reports (see Attachments; Conclusions; Discussion; Introduction; Recommendations; Summary)

182 Index

Reports (cont.)
 production of reports, 173–78
  illustrator, working with, 177
  management editing, 177–78
  printer, working with, 177
  schedule, 174
  team approach, 173–74
  typists, working with, 175
  word processor, working with, 176–77
 types of reports:
  formal report, 99–130
  inspection report, 26–33
  investigation report, 45–47, 63–75, 99–130
  laboratory report, 51–61
  occurrence report, 17–19
  progress report, 35–45
  trip report, 20–26
 typing guidelines, 138
 writing reports, 139–49 (see also Writing techniques)
Reporting conference attendance, 26 (see also Trip report)

Summary (see also Summary Statements):
 in formal proposal, 94–95
 in formal report, 99, 100, 101, 103, 115
 in investigation report, 64, 65, 69
 in laboratory report:
  academic lab report, 60, 61
  industrial lab report, 52
 as main message, 14
 in semiformal proposal, 80, 81, 82, 86
Summary Statement (see also Summary):
 basic application to reports, 10–14
 in inspection reports, 26, 28, 30
 in investigation report, short, 45, 46
 as main message, 10–12
 in occurrence report, 18
 in progress reports, 36, 37, 38, 40, 42
 in trip reports, 20, 21, 24
S.I. units, guidelines for writing, 171
Site plans, 166
Spelling, 169–70
Starting writing, 5–6, 139–40
Suggestion, informal, 77, 78–80 (see also Proposal)
Supervisory editing, 177–78
Surface chart, 164–65
Synopsis, 42 (see also Summary Statement)

Table of contents, formal report, 100, 101, 103–4, 116
Tables, 158–59
Test report, 51–61 (see also Laboratory report)
Title page, formal report, 100, 101, 102, 114
Trip report, 20–26
 application to conference attendance, 26
 circumstances for writing, 20
 comments on examples, 21–22, 24
 examples, 23, 25
 headings, typical, 21
 reasons for trip reports, 20
 writing compartments, 20
Typists, working with, 175

Voice, active vs passive, 145–47

Word processors, working with, 175–76
Wordy expressions, eliminating, 147–79
Writing:
 climatic method, 9
 getting started, 5–6
 message, identification of, 7
 organization method,
  basic writing compartments, 13–14
  example, (fig.) 14
 plan, semiformal proposal, 80–81, 84–85
 purpose, identification of, 140–41
 pyramid method, 6–14 (see also Pyramid method of writing)
 reader, need to identify, 6–7, 140
 techniques for writing, 139–49
  abbreviations, forming, 170
  active voice, preference for, 145–47
  being direct, 141–47
  clichés, removing, 148, 149
  first person, use of, 143–44
  getting started, 5–6, 139–40
  informative writing, 141
  metric units, using, 171
  numbers, using in narrative, 171
  passive voice, use of, 145–47
  persuasive writing, 141
  pyramid writing, 142–43
  S.I. units, guidelines for using, 171
  spelling, 169–70
  word choice, 147
  wordy expressions, removing, 147–49
 typists, working with, 175
 word processors, working with, 175–76